BUCKNELL REVIEW

Lorca, Buñuel, Dalí: Art and Theory

STATEMENT OF POLICY

BUCKNELL REVIEW is a scholarly interdisciplinary journal. Each issue is devoted to a major theme or movement in the humanities or sciences, or to two or three closely related topics. The editors invite heterodox, orthodox, and speculative ideas and welcome manuscripts from any enterprising scholar in the humanities and sciences.

This journal is a member of the Conference of Editors of Learned Journals

BUCKNELL REVIEW
A Scholarly Journal of Letters, Arts, and Sciences

Contributors should send manuscripts with a self-addressed stamped envelope to the Editor, *Bucknell Review,* Bucknell University, Lewisburg, PA, 17837.

BUCKNELL REVIEW

Lorca, Buñuel, Dalí: Art and Theory

Edited by
MANUEL DELGADO MORALES
and
ALICE J. POUST

Lewisburg
Bucknell University Press
London and Toronto: Associated University Presses

Associated University Presses
440 Forsgate Drive
Cranbury, NJ 08512

Associated University Presses
16 Barter Street
London WC1A 2AH, England

Associated University Presses
P.O. Box 338, Port Credit
Mississauga, Ontario
Canada L5G 4L8

The paper used in this publication meets the
requirements of the American National Standards for
Permanence of Paper for Printed Library Materials Z39.48-1984.

(Volume XLV, Number 1)

ISBN 0-8387-5508-9
ISSN 0007-2869

PRINTED IN THE UNITED STATES OF AMERICA

Contents

Recent Issues of BUCKNELL REVIEW

Mappings of the Biblical Terrain: The Bible as Text
The Philosophy of John William Miller
Culture and Education in Victorian England
Classics and Cinema
Reconfiguring the Renaissance: Essays in Critical Materialism
Wordsworth in Context
Turning the Century: Feminist Theory in the 1990s
Black/White Writing: Essays on South African Literature
Worldviews and Ecology
Irishness and (Post)Modernism
Anthropology and the German Enlightenment: Perspectives on Humanity
Having Our Way: Women Rewriting Tradition in Twentieth-Century America
Self-Conscious Art: A Tribute to John W. Kronik
Sound and Light: La Monte Young/Marian Zazeela
Perspectives on Contemporary Spanish American Theatre
Reviewing Orpheus: Essays on the Cinema and Art of Jean Cocteau
Questioning History: The Postmodern Turn to the Eighteenth Century
Making History: Textuality and the Forms of Eighteenth-Century Culture
History and Memory: Suffering and Art
Rites of Passage in Ancient Greece: Literature, Religion, Society
Bakhtin and the Nation
New Essays in Ecofeminist Literary Criticism
Caribbean Cultural Identities

Notes on Contributors

MIRIAM BALBOA ECHEVERRÍA is a professor of Spanish at Southwest Texas State University. She is the author of *Lorca: El espacio de la representación* (1986), *Boca de dama: La narrativa de Angélica Gorodischer* (1995), a play, *Doña Catalina* (1996), and a book of poetry, *Para entretener a un amante* (1999). Her articles have appeared in *Gestos, Romanic Review, Discurso Literario, Mesoamérica, Confluencia,* and *Unesco.*

MANUEL DELGADO MORALES is a professor of Spanish at Bucknell University. A native of Granada, Spain, he is the author of a number of books and articles on the theater of Gil Vicente, Lope de Vega, Guillén de Castro, Calderón de la Barca, Tirso de Molina, Ruiz de Alarcón, Mira de Amescua, and Federico García Lorca. His most recent publication is a critical edition of Pedro Calderón de la Barca's *La devoción de la cruz* (2000).

SIDNEY DONNELL is an associate professor of Spanish at Lafayette College. His forthcoming book *Feminizing the Enemy: Imperial Spain, Transvestite Drama, and the Crisis of Masculinity* looks at gender politics in the early modern period by focusing on the deployment of cross-dressing in the Spanish *comedia.* He also researches film, having recently published an article in *Modern Language Notes* about Luis Buñuel.

LUIS FERNÁNDEZ CIFUENTES is a professor of Spanish at Harvard University. He is the author of numerous books and articles, among them *Teoría y mercado de la novela en España* (1983) and *García Lorca en el teatro: La norma y la diferencia* (1986). He is also the editor of Zorrilla' s *Don Juan Tenorio* (1993), Palacio Valdés's *Los majos de Cádiz* (1998), and Max Aub's *Las buenas intenciones* and *La calle de Valverde* (forthcoming).

CANDELAS GALA is a professor and Chair of the Department of Romance Languages at Wake Forest University. A native of Santander,

Spain, she is the author of three books on Lorca, and editor of a volume of essays on Spanish, Latin American, and US/Latino women authors. She has also published numerous essays on works by twentieth-century poets, particularly by women poets, such as Clara Janés, Pureza Canelo, Cristina Lacasa, María Beneyto, María Victoria Atencia, Carmen Conde, Rosa Chacel, Gloria Fuertes, and Concha Lagos, among others.

VIRGINIA HIGGINBOTHAM is an associate professor of Spanish at the University of Texas, Austin. She is the author of books on Lorca, Buñuel, and a number of studies on modern Spanish film, the most recent, a monograph on *Spirit of the Beehive.* She is currently preparing a study of women's writing in postwar Spain.

MICHAEL IAROCCI is an assistant professor of Spanish at the University of California, Berkeley, where he teaches courses on modern Spanish literature (eighteenth–twentieth centuries). He is the author of *Enrique Gil y Carrasco y la genealogía de la lírica moderna* (1999) and has published articles on the work of authors ranging from Cadalso, to Larra, to Pereda. He is editor of the monograph series, "Hispanic Enlightenment," recently launched by Bucknell University Press. His current research is on the aesthetic ideologies of Spanish Romanticism.

ROBERT LIMA is a professor of Spanish and comparative literature and Fellow of the Institute for the Arts and Humanistic Studies at The Pennsylvania State University. He is the author of numerous books and articles, among them *The Theatre of García Lorca,* the first critical study in English to address the plays. He is the biographer of Valle Inclán in English and translator of several of his plays and *The Lamp of Marvels.* He has published extensively on Borges and, most recently, *Dark Prisms: Occultism in Hispanic Drama* and *Ramón del Valle Inclán: An Annotated Bibliography* (3 vols.).

ANTONIO MONEGAL is a professor of literary theory, comparative literature and film at the Universitat Pompeu Fabra (Barcelona). He is the author of *Luis Buñuel de la literatura al cine: Una poética del objeto* (1993) and *En los límites de la diferencia: Poesía e imagen en las vanguardias hispánicas* (1998). He has also edited Lorca's *El público* (2000) and *Viaje a la luna* (1994) and cooordinated the co-authored volumes *Sin fronteras: Ensayos de literatura comparada en homenaje a Claudio Guillén* (1999), *Federico García Lorca i Catalunya* (2000), and *Literatura y pintura* (2000).

PITHAMBER R. POLSANI is an assistant professor at the University of Arizona, Tucson. His research interests are in Spanish Golden Age drama, twentieth-century European philosophy and cultural studies. He is currently working on a multimedia edition of Lope de Vega's *Fuenteovejuna,* and a book-length manuscript that explores how the concept of representation in philosophy in the seventeenth century occasions the rise of Spanish baroque theater.

ALICE J. POUST is an associate professor of Spanish and Chair of the Spanish department at Bucknell University. Her research deals with twentieth-century Spanish and Spanish American literature, primarily in the poetic genre. She has published articles on pre- and post-civil war Spanish poetry and aesthetics as well as on Spanish American poetry in such journals as *Cuadernos Hispano-americanos, Hispania, Symposium, Revista de Estudios Hispánicos, Revista Hispánica Moderna,* and *Revista Canadiense de Estudios Hispánicos.*

C. CHRISTOPHER SOUFAS JR. is a professor of Spanish at Tulane University. He is the author of *Conflict of Light and Wind: The Spanish Generation of 1927 and the Ideology of Poetic Form* and *Audience and Authority in the Modernist Theatre of Federico García Lorca.* He has also edited three volumes on contemporary theory, postwar novel, and a collection on interdisciplinary/interartistic studies. He is the author of numerous articles that have appeared in *PMLA, Modern Language Notes, Hispanic Review, Bulletin of Hispanic Studies, Word and Image, Gestos,* among others.

Introduction

THE artistic creations and personal lives of Federico García Lorca, Luis Buñuel, and Salvador Dalí, three of the most celebrated Spanish artists of the twentieth century, have elicited remarkable and sustained critical interest in Spain and throughout the world. The centenary of their births, 1998 for Lorca, 2000 for Buñuel, and 2004 for Dalí, has provided the impulse for new reflections on their works and lives as individuals and also on their relationship with one another, as well as on possible links to other artists and intellectuals of their day. The scholars featured in this issue of the *Bucknell Review* are pleased to join with others throughout the world in honoring these now iconic figures who seem in many respects to be our contemporaries.

As is well known, Buñuel, Lorca, and Dalí lived together for a time at the *Residencia de Estudiantes* in Madrid, a prestigious cultural institution designed to foster and shape the intellectual development of Spain's brightest young thinkers, writers, and artists, from its foundation in 1910 until the outbreak of the Spanish civil war in 1936. At the *Residencia de Estudiantes* these three, then young, men were exposed to the ideas that were already shaping the intellectual and artistic life of Spain and Europe. The personages who visited and often lectured at the *Residencia* included José Ortega y Gasset, José Martínez Ruiz (Azorín), Miguel de Unamuno, Ramón Menéndez Pidal, Ramón del Valle Inclán, Andrés Segovia, Manuel de Falla, Maurice Ravel, Marie Curie, Albert Einstein, Paul Valéry, Paul Claudel, and Louis Aragon. As this list suggests, the young *residentes* were exposed to some of the most authoritative thinkers of their day from a wide variety of intellectual endeavors.

Lorca's violent death in the early days of the Spanish civil war served to imprint his memory on his own and subsequent generations of Hispanic nations, while the intellectual diaspora produced by the civil war helped to make the works of Buñuel and Dalí known far beyond the national borders. At the same time, it is clear that each of these men, while keenly conscious of their roots in their particular regions of Spain, sought creative inspiration and experiences in an international arena. The present volume comprises critical studies that focus on the ways in which the works of Lorca, Dalí, and Buñuel seem to reach out, even allowing for a certain characteristic hermeticism shared by all three, to initiate a dialogue with other artists and intellectuals of their day and beyond. The essays featured here generally focus on what was new and even revolutionary about their work from the perspectives and times of Lorca, Dalí, and Buñuel, but, also what has become a familiar cultural landmark for subsequent generations of artists and critics. In addition, these essays help to situate the works of these three Spanish figures within a broader context of contemporary aesthetic as well as cultural, social, psychological theory.

Miriam Balboa Echeverría focuses on the figures of the female servant in the plays of Lorca in "The Point of Desire and Death: Reading the Maids in Lorca," highlighting, on the one hand, the differences that separate Lorca's use of this character from that of Golden Age theater, and on the other, the similarities between Lorca's dramatic use of the female servant character and that of the twentieth-century French dramatist Jean Genet. Echeverría shows how Lorca's use of the female servant's socially marginal status in his plays prefigures that of Genet in its tendency to both undermine and imitate the social repression on which upper-class honor and privilege are based.

In "Embroiderers of Freedom and Desire in Lorca's Poetry and Theater," Manuel Delgado Morales studies the portrayal of women engaged in needlework in the work of Lorca. Delgado argues that Lorca's discovery of Jan Vermeer's *Lacemaker* in the *Residencia de Estudiantes,* a discovery shared by Dalí and Buñuel, would become, through the mediation of Freud's interpretative theory, a powerful symbol of sexual desire and liberation. Delgado sees in Lorca's portrayal of this handiwork an attempt to subvert the social, political, sexual, and psychological constraints that prevent them from fulfilling their desires and aspirations.

In his essay "If the Shoe Fits: Buñuel, Almodóvar, and the Modernist/Postmodernist Question," Sidney Donnell juxtaposes films by Buñuel and Pedro Almodóvar to show how Almodóvar pays homage to Buñuel by quoting several scenes from the latter's comic film noir *Ensayo de un crimen* (Rehearsal of a crime/The criminal life of Archibaldo de la Cruz) in his recent motion picture *Carne trémula* (Live flesh). Donnell concludes that while Buñuel and Almodóvar share certain characteristics with a postmodern aesthetic vision, their respective works are, to varying degrees, historically and politically invested art forms within the context of modernity.

In "1918: García Lorca at the Crossroads," Luis Fernández Cifuentes considers Lorca's first book (and only travelogue), *Impresiones y paisajes,* to be a generational turning point. While the genre and subject matter pay tribute to the cartography of the older Generation of 1898, the youthful work manifests a resistance to the older generation's preference for the country over the city, for the ancient and traditional over the modern, and for Castile as a model for the rest of the nation. This early Lorquian work reflects a new sensibility that will favor a more flexible, heterogeneous, protean national and international map.

Taking parodic *guasa* as the overriding rhetorical device, Candelas Gala explores Lorca's personal and artistic concerns at the time he wrote the seven poems in the section "Eros with a Cane" from his book *Canciones* (Songs). Gala affirms in this study, " 'Tópicos sublimados': Lorca's Female Iconographies in 'Eros con bastón,' " that these poems offer a kind of portrait gallery of female prototypes well known in popular iconography. She shows how Lorca pokes fun at stereotypical attitudes about sex and artistic anxiety through the use of such devices as the Gongorine aesthetics of the gaze, self-referentiality, self-reflection, and intertextuality, all related to parodic *guasa,* devices that also exhibit an artistic awareness of modernism.

In "Luis Buñuel as Intertext in the Narrative Prose of Juan Goytisolo," Virginia Higginbotham reflects on the fact that film, once considered a frivolous and impermanent medium, can have as lasting an impact as any of the plastic arts. She uses Roland Barthes's notion of intertextuality as a "plurality of other texts" and Julia Kristeva's discussion of the intersection of texts to demonstrate how Buñuel's use of surrealism as a system of signification and film as a narrative device in his films still resonate in Spain today, particularly in the late prose of one of Spain's leading writers, Juan Goytisolo.

Michael Iarocci analyzes the role of neoromantic rhetoric in Lorca's *Libro de poemas,* arguing that this Lorquian work thematizes a break with Romantic metaphysics. In "Romanticism, Transcendence, and Modernity in Lorca's *Libro de poemas,* or the Adventures of a Snail," he analyzes this break within a number of interpretive frames: as psychobiography it marks the lyric subject's transition to adulthood; in spiritual terms, it bespeaks a loss of faith; metapoetically, it registers the unraveling of quasi-allegorical representation. Finally, within the broader framework of cultural history, *Libro de poemas* can be read as an allegory of the processes of secularization at work in Spain's early twentieth century.

In "Missing in Action: Invisible Males in *La casa de Bernarda Alba,*" Robert Lima focuses his attention on Lorca's tendency to have heterosexual male characters as relatively secondary figures who, moreover, often die in his works as a result of their romantic involvement with women. Lima cites the examples of Don Perlimplin in the work by that name, Leonardo and El Novio in *Bodas de sangre* (*Blood Wedding*), and El Joven in *Asi que pasen cinco años* (*Until five years pass*), among others, noting that in *La casa de Bernarda Alba* (*The House of Bernarda Alba*) men are barred from the stage altogether. Based upon this tendency, Lima conjectures that Lorca might have eliminated heterosexual males completely from his dramaturgy had he lived to write more works.

In his study "Shall the Circle Be Unbroken? Verbal and Visual Poetry in Lorca, Buñuel, and Dalí," Antonio Monegal explores the ideas and vision of each with regard to his own art form, and on the interaction among the arts in the avant-garde. Notwithstanding the general interpenetration of poetry, painting, and film in avant-garde movements of their time, Monegal affirms, Lorca, Dalí, and Buñuel made poetry the common denominator for artistic creation in general. For all three, literary creation, painting, filmmaking, were poetic, and they tended to use poetry as a system for constructing meaning through the juxtaposition of signs and silences as a creative model for their works.

Pithamber R. Polsani analyzes the role of a theoretical paradigm called Paranoiac-Critical Activity, developed by Dalí himself in the artist's works from the 1930s, in this essay entitled "The Image in a Fatal Kiss: Dalí, Lacan, and the Paranoiac Representation." He outlines the application of this method, in which the painter (Dalí) would paint the image that comes to his mind first, and add other images that are suggested by the first one, noting that the paranoiac method is not a random grouping of disparate images but a system in itself. Polsani demonstrates how each of the elements of Dalí's Paranoiac-Critical Activity relates to Jacques Lacan's reading of paranoiac delirium, as well as how Dalí achieves his goal of creating systematic confusion in reality.

Alice J. Poust examines the cultural theory implicit in Lorca's systematic use of Andalusia as a symbolic space for his poetry and theater in "Federico García Lorca's Andalusia in Light of Oswald Spengler's Theory of Magian Cultures." She traces parallels between the cultural ideas modeled by Lorca's work and the theory of culture formulated by Spengler in *The Decline of the West.* Focusing on Lorca's views on several facets of Andalusian culture (ethnicity, the symbolism of the cave/dome structure, the image of the arabesque, *duende,* and alchemy) in light of Spengler's theory of cultural history, she suggests a new theoretical framework for Lorca's ideas on culture.

In "Lorca beside Artaud: Parallel Developments of a Modernist Mise-en-Scène," C. Christopher Soufas Jr. presents a comparative study of the theatrical theories and practices of Lorca and his contemporary Antonin Artaud in order to underscore not only their remarkable commonality of thinking on the reform of the European stage, but also to suggest that Lorca has not been given enough consideration as a thinker and theorist regarding the early twentieth-century stage. Soufas finds that the similarity of Lorca's ideas regarding the theater as mise-en-scène to those of Artaud, long recognized for his innovations in performance theory, support the consideration of Lorca as an equal among his European contemporaries as both a playwright and an innovator.

MANUEL DELGADO MORALES
ALICE J. POUST

BUCKNELL REVIEW

Lorca, Buñuel, Dalí: Art and Theory

The Point of Desire and Death: Reading the Maids in Lorca

Miriam Balboa Echeverría
Southwest Texas State University

SURREALISM is the breaking point of Lorca's theater, the point at which the form, techniques, and themes of his earlier theater undergo a devastating fragmentation, providing the basic aesthetic principles for the artistic translation of a fragmented modern consciousness. This inner fragmentation leads to a theater played out in an inner mental space performed in the suffocating square form of *The House of Bernarda Alba,* in which the walls of the house are like the frames of black and white negatives, one on top of the other, and whose margins are dangerously phasing out in constant movement, blurring photographic reality, suggesting madness.[1]

In the context of the European avant-garde, Lorca's work is the juncture between the unity of baroque performance texts and the texts of a fragmented modern mind. The following study seeks to study this point of juncture articulated in a corpus of so-called minor characters that requires an interpretative framework for its foregrounding of cultural control, transgression, and cruelty. In the margins of the texts and the margins of the culture the maids and other minor characters constitute a particularly relevant case for studying patterns of cultural hatred and have been studied insufficiently to date. Centering on the surrealist theme of motion and immobility inherent to Lorca's work, a number of movements and theories pertaining to painting, photography, and theatrical performance will be examined in order to better understand Lorca's critique of his culture.

Lorca's surrealist work with its evident parallelism with Artaud's "theater of cruelty," anticipates later theatrical constructs, such as travestism, infrarealities, slow human decay, and antimimetic theater, and announces directions to be taken by Becket, Ionesco, and Genet. In *The Public,* which precedes Genet's theater, the characters play the roles of actors and perform transvestite games in multiple rituals of self-conscious transformations which change previous perceptions of illusion. In one scene the characters of Romeo and Juliet are played by a man and a boy while the actress who should play Juliet is gagged under a seat in the audience; in another scene the dialogue between the Character in Vine Leaves and the Character in Bells assaults the spectator with its allusions to fecal material, with forms that pullulate—funguses, caca, spiders, excrement, ants, earth, flies, and wheat. The constant verbal game plays out a convoluted performance in the form of a theatrical reality that turns into sequences of illusions, desire, and terror:

> *Figure in Bells.* If I turn into a cloud?
> *Figure in Vine Leaves.* I'd turn into an eye.
> *Figure in Bells.* If I turn into caca?
> *Figure in Vine Leaves.* I'd turn into a fly.
> *Figure in Bells.* If I turn into an apple?
> *Figure in Vine Leaves.* I'd turn into a kiss.
> *Figure in Bells.* If a turned into a breast?
> *Figure in Vine Leaves.* I'd turn into a white sheet.
> *Voice* [sarcastically]. Bravo!
> *Figure in Bells.* And if I turned into a moon-fish?
> *Figure in Vine Leaves.* I'd turn into a knife.[2]

In this play Lorca is not concerned with a theater that stages the control of power in the way of mimetic theater. Instead, his point of departure is from human illusion trapped dialectically in a framed space for the scrutiny of others, thus a theatrical reality that keeps transforming itself into another searching for freedom by its successive accumulation of appearances. But the self-reflecting vision that confronts itself will ultimately be betrayed since the characters are aware of their own and each other's infinite representation. This changes radically the established relationship between audience, actors, and theater, creating a complex montage of desire and terror shared by an audience who perceives itself in danger as it looks through the double vision of mirrors.[3] Thus Lorca is creating a theater that is moving toward the emergence of a modern consciousness, much like that of Jean Genet, who will seek to make the inner mental spaces of the audience a part of the spectacle through what Oreste Pucciani has called an "inversion of theater," in effect turning the performance into a "black mass."[4]

The metamorphosis of Lorca's characters in this play depends on the surrealist clash produced by the violent juxtaposition of the self with the Other and seems to announce the complex tensions of hatred and closure in Genet's *The Maids* and in *The House of Bernarda*

Alba. In the Roman Ruin scene, the Character in Vine Leaves is clearly a reflection of the Red Nude character in the fifth scene, who is crowned with blue thorns, imaging pagan Rome and Christian crucifixion. Furthermore the Emperor is transformed later into Pontius Pilate, accentuating the continuous desire to be Other and to escape. These constant transmutations have a historical background: the unstable location where pagan Rome transforms itself into Christian civilization in a constant crossing of limits. Lorca's modern game of transgressions transforms the Epicurean crown into a Christian one, creating dual perspectives for the spectator, who had been, until now, seated in the audience of a mimetic theater watching from a fixed point in reality.

Lorca's surrealist drama was off-putting and shocking for the audience of the period; as Lorca himself stated, according to Mildred Adams, "this is for the theater years from now."[5] Lorca's plays were written the decade preceding the Spanish civil war and the Second World War. Genet's theater appears after the war, confronting an audience that desired a fixed reality for the stage, an audience that resisted the ritualizing of terror. The violence of Lorca's surrealist theatrical discourse announces, as I indicate above, the maids in *The House of Bernarda Alba* and the terrorizing maids of Jean Genet. These characters are not usually considered central in the dramatic representation; thus their space has absolute potential for expressing desires against the grain, while the framed constructs of modern discourses hiding a fragmented psyche are curiously disassembled and reassembled by these marginal figures. Both in the baroque and in the modern era, this corpus of women operates within a cultural articulation that requires an interpretative framework for the study of culture, which this essay will address. While in baroque theater the role of the maid subverts the status quo, in modern drama, as we shall see, it ritualizes terror by moving between ideological and sociological coordinates.

Baroque art, with its interplay of multiple perspectives and mirrors reflecting into each other a world of appearances, regimented life and was constantly challenged from intricate spaces created by the invention of complex systems that permitted the subject a space for "letting off steam."[6] As José Antonio Maravall has stated, in seventeenth-century Spain there were very elaborated holiday celebrations that served a political function. Stage representations were part of the "divertissement" funded by the monarchy to produce a controlled theater that safeguarded the social order and at the same time served as an escape from the regimented life imposed on the individual.[7] While theater served as a diversion, these afternoons of performances and carnivals were not just to deflect the gaze from the king's power. In the performance itself there was a marginal space that permitted spectators and players a reflexive space to scrutinize the social structure. It centered the gaze on the "in between margins" if we follow Victor Turner's explanation of cultural frames.[8]

From the baroque period, I scrutinize specifically the figure of Maribárbola, whose body posture at the threshold of the painting alters the perspective of the gaze in Velázquez's *Las Meninas.* From Calderón de la Barca's play *A House of Two Doors Is Difficult to Guard,* I analyze the figure of the maid Celia, who mocks the cultural system not only by parodying discourses but by positioning the body at the edge of the stage and by the grain of her voice.

At first glance the painting *Las Meninas* shows that life and social milieus were perfectly ordered according to the official hierarchy of the court; each one of the figures depicted played a specific role assigned to him or her according to social class and daily performance. As discussed by Foucault, the painting is a tableau designed to make the spectator and the painter the center of the performance since the royal presence is only a reflection in the mirror.[9] For his part, Kenneth Clarke has emphasized Maribárbola as a very strong presence: "she affronts the spectator like a blow,"[10] for she is gazing directly at the king watching the sitting and at the spectators always present in front of the picture. I would suggest, then, that Maribárbola, the painter, and the spectators are out of the hierarchy but at the center of the performance. Although the subject is ostensibly ladies-in-waiting, Velázquez has rearranged the tableau by using gaze intersections, space and light interplay, and, most important, by positioning bodies as if on a stage, to emphasize a different perspective on the regulated order.

In another perfect, immobile tableau, at center stage, the group of the two young ladies-in-waiting and the princess form a semi-circle. Their bodies, with very precisely sketched lines, colors, and soft lights, have the meticulous detailing of porcelain figurines frozen in a still, balletlike pose to be gazed at and admired, for they represent the world of appearances. The two young ladies' faces are only profiles in a triangular flat dimension without depth. The princess's face is totally illuminated as if by a camera flash, yet the perfect face remains aloof from the spectator. It does not commit itself to a dialogue as Maribárbola does with us.

If the central focus of the baroque painting was perspective and the perception of existing in a world of appearances, theater is the performance of existing in an unstable mental space. In his painting

Velázquez changes for the spectator the perceptions of the world of appearances, while in his play Calderón reverses the balance of hierarchies. The reflexive space is generated when the actors address the characters' discourse directly to the spectator using the theatrical device of the aside, thus creating a place in between the margins of the official and the social. The aside permitted the actors to communicate a double entendre and the author was able to create a reflexive space for the spectators, who then could observe, with an ironical gaze and from an unstable place in reality, the game being played on stage, the game of honor. And this brings me to my central point about the maids: In *A House with Two Doors Is Difficult to Guard,* Calderón uses as the point of attack the maids, who through the convention of the aside make comments about the private designs of their veiled ladies transgressing in comic relief the rules of the game and becoming a central force in the play.

As I have suggested before in another work,[11] in the baroque comedies of veiled ladies and suitors, the maids not only initiate technically the point of attack and the closure of acts by playing with double meaning, but also initiate with their ladies a web of interconnecting labyrinths representing woman's writing and desire. In *A House with Two Doors* Calderón manipulates the social order by using a maid to initiate the dramatic movement and rhythm of the play to involve the spectators in a farcical game of honor. The maids Celia and Silvia perform from the stage margins, crossing its limits toward the spectators and delivering their lines in a double play-acting as ironical commentators of the other characters' games, opening the game from the inside out. The maids constantly cross the spaces in between the houses, constructing narrative passageways as laughing accomplices of their veiled ladies' serpentine webs. Simultaneously, they are spectators of the upper class's amorous games and are in charge of closure at the ends of acts as they parody the main action of the plot. The maids and their ladies, playing their own games in private spaces behind closed doors, write a discourse under the surface that contrasts with the patriarchal façade, thus permitting them to accomplish their desires and subverting the masculine order.

In Velázquez's baroque painting the ironical yet direct gaze of Mirabárbola decentered the equilibrium of the hierarchy. Moreover, by using light, body formations, separated perspectives and volumes, the painter has created a textural weaving, transforming the atmosphere of court appearances and making us, the spectators, reflect on our own unstable condition. On the baroque stage the maids take the sense of instability to the extreme and ultimately subvert the order of appearances. One of the maids, Celia, constructs the closure of act 1, laughing at the lovers' discourse and, by implication, laughing at the probably absent royal presence playing with double entendre the reference to honor. Repeating in jest the lovers' laments, Celia stretches her body and voice toward the audience and unbalances the social conventions reaching out for that inner space of the spectator's mind and by adding the comment about the double meaning of the doors through her ironical wink at the audience: "*Celia:* Ah! how convenient / It is to have a house that has two doors!"[12]

The doors mark the crossing of thresholds and facilitate the comings and goings of the suitors, making the audience laugh at the positioning of Celia's body in between stage and audience, and at her voice echoing the tone of the lovers' discourses emphasizing "Ah! how convenient," mocking the seriousness of the amorous grievances. The double meaning of a house with two doors signals not only the comical gesturing of a mischievous maid but a more dangerous game in which the spectators are involved since they know the rules of the game. Neither the spectators nor the author can transgress the public space; it is only through the grain of the voice of a servant girl that the author is able to communicate naughtily the space transgressions and the uncovering for a brief moment of the hypocritical public discourse that imprisons the subject in the regimented social space and imprisons women behind the closed doors of a house.

Velázquez's baroque painting and the Calderonian stage then use the light to seize the gaze, focus the awareness of being in an oblique way to deform the architectural formation of bodies. These stops in texture are designed to signal the performance becoming a series of performances, as in blurred negatives creating a reflexive inner space for the spectator. As the gaze illuminates the private space through Maribárbola, and the constructions of theater decenter the official equilibrium through Celia, both as intertwinings of private and official, it is still possible for the marginal subject to look through the opening inventing its own space. However, in modern theater all possible openings are closed like proliferating but shattered cubic mirrors, one on top of the other with different irregular edges, exacerbating the closure of the social structures' previous passageways.

These works allowed for elements that threatened to subvert the status quo, yet contained those elements within a framework that sustained the status quo. By contrast, the texts of Genet and Lorca constitute a rejection of a mimetic theater that represents the control of power, on the one hand, and that constructs a cultural epistemology, on the other. Their texts reflect ironically on how history is codified to legitimize Western thought. This reflection is constructed through a perverse game with similes imposed by traditional philosophical discourse. In this game the similes appear as the carnivalesque locus of logocentric thought. Its main characteristic seems to be the consciousness of performance and the staging of an accumulation of cultural frames.[13]

According to Turner, cultural frames scrutinize segments of the social, isolating them and intensifying the sense of ritual.[14] In Genet's case the theatrical locus is always ritualized as an inside double spectacle for the characters who frame rituals, one inside the other, as if the spectacle were the performance of the performance. From the outside in, the repressed classes perversely play a game of metamorphosing themselves into the oppressors with secret rituals that are self-referential and autodestructive.

The rituals become the game, a never-ending game of ritualization. They obsessively create frames inside one another until the game cannot be stopped and it becomes the prison, the terror. The characters, as actors playing the oppressors, are aware of playing at being actors, but they are immobilized by the repressive social system that controls and seduces them into playing on. Playing the oppressor generates for the characters what Jameson calls a desire and admiration for the political apparatus that is excluding them.[15]

In *The Maids* the consciousness of representation clearly announces the sense of terror. Claire says "the game is dangerous,"[16] referring to the fact that, if the characters representing power discover the secret game, the game will no longer be the maids' game, but a ritual of the oppressors. Nevertheless the game itself is dangerous because it is the point of desire. At an epistemological level the ritual is dangerous because even though it stages the hate and rejection of the oppressors, it simultaneously perpetuates the repressive system by transforming it into rites that lock in the agents of power in a suffocating repetition without exit, betraying the intent to denounce the power structures.

In Lorca's play *The Public,* as I mention above, the similar attempts by the characters to free themselves by constant metamorphosis result in the affirmation of power, that of the Emperor and the public, making apparent to the audience the emergence of a sense of terror when confronting repressive structures that were also seductive to the oppressed. The choices made by the maids reflect the attraction to their own ultimate destruction since, according to Silvio Gaggi, the ritual neither liberates nor is cathartic.[17] On the contrary, they generate the self-referential terror of nonbeing, the nothingness of a "toilet-seat":

> *Solange:* She loves us the way she loves her armchair. Not even that much! Like her bidet, rather. Like her pink enamel toilet-seat. And we, we can't love each other . . . Filth doesn't love filth . . . It is easy to be kind, and smiling, and sweet—ah! that sweetness of hers!—when you are beautiful and rich . . . But what if you're a maid? (*M*, 52)

In *The Maids* rituals not only serve to officiate the hatred for Madame, but also for Claire and Solange, since their lives are hate performances that take place in private spaces inside the house and always in the margins. The roles on the first level of performance had been defined for them by the masters; the essence of playing at being maids existed before the play in the articulation of the culture. On top of this emphatic layer of inert performing, they shift the gaze to the writing of their own scripts with their own desires, as they perform rituals of continuous metamorphosis, always playing Madame and themselves as maids. Not only are the stagings private, framed and closed, but the mental spaces are molecularized in the Lacanian sense, which means Claire and Solange can only play their roles if they center themselves in a specific mental space that will separate them from their real being.

When playing the role of maids, Claire and Solange hate their masters. However, when Claire plays Madame and Solange plays Claire in what Martin Esslin describes as "a hall of mirrors" and Sartre as a "whirligig of appearances,"[18] they succumb to the seduction of the game and hate each other. The continuous rites have phased out the precise cultural frames of representation; it is not a frame inside another or a continuous construction of successive frames but endless fading frames revealing simultaneous negatives of other frames with blurred limits. The mere representation of power is not the center of the spectator's gaze but the character's body with a furious desire for its political right as subject.

Power is produced by the oppressors who control it and stage it to make it visible; these oppressors do not need to be there in the mise-en-scène to exercise it. In *The Maids* the house is the place of power that will be perpetuated by Claire and Solange. The rituals will involve the audience in a complicity of terror, making it impossible for the maids to control the proliferation of rituals which ultimately will be their destruction.

The character of Madame is characterized by Genet as a puppet of the system, an iconic sign of power, play-acting in front of her maids: "Of course, none of this is serious but if it were, Solange, it would be a joy for me to bear his cross" (*M*, 67). Madame is "wild with happiness" (*M*, 74) when she hears Monsieur is out of prison, although she knows she is being hypocritical, for his imprisonment was to be short, as the system protects those who represent it. This is revealed when Monsieur is released from prison and Solange exclaims, "The judges have their damned nerve letting him go. It's a mockery of justice. It's an insult to us!"[19] Solange is contemplating herself in the mirror caught up in the double realization of the interconnection between dependency and power and of "the impossibility of revolution in Western society, as Goldmann has affirmed."[20]

The iterative game of officiating framed rituals becomes more and more asphyxiating for Solange and Claire, representing its futility and the failure of institutions to obtain legal status for the oppressed. The only solution left is that of the ultimate game, playing at not playing; although Claire had declared early in the play "the game is dangerous," now the game will be to play their last ritual in an arabesque change of roles as they pretend to poison Madame, played by Claire. As a result of Solange poisoning Claire, Solange will become more than a maid, being singled out as the murderer "mademoiselle Solange Lemercier, that Lemercier woman. The famous criminal" (*M*, 95). Claire has decided to be killed only to give an identifiable image to Solange, the only kind of recognition available for a maid from an institution of power.

The play ends with an incantation in which Claire as Madame forces Solange to give her the poisoned tea; while she drinks, Solange delivers the last lines to the audience narrating a hypothetical grandiose spectacle of Madame as a body executing daily activities, but as a body that is dead while the maids are alive "raised from Madame's icy form" (*M*, 100). She talks of "all the maids" (*M*, 99) as a virtual corporal form that is finally free in a narrative tone changing the last line in which she finally includes herself: "We are beautiful, joyous, drunk and free" (*M*, 100). The real freedom of the maids as an entity is only theoretical since its condition of being is a death wish, which in turn reflects on the audience as a metaphor of society playing against itself.

In this play Genet asked for a declamatory tone of acting "Her gestures—arm extended—and tone are exaggeratedly tragic" (*M*, 35) that would show the displacement between actor and characters clearly playing a profusion of roles against the audience and against theater itself.[21] Playing performance over series of performances creates simulacra that cross between the audience's inner mental space and the representation, liquidating the real and creating a violence recurring over and over.[22] The distance between actor, spectator, and characters stages a series of superimposed spaces which, by concealing an emptiness of meaning, projects multiple representations, including the sign of asphyxiation produced by the violence of repressive institutions.

Genet's maids and their games, stretching the limits of artificial representation and constant polarization for the control of power, are announced in Lorca's configuration of his maids and other minor characters which will project onto the audience the idea of terror, for the gaze is centered on the performance of the cracks and interstices.

The repressed subjects, in this case the figure of the maid, use violence and evil to overcome the terror of the suffocating cultural structures surrounding them, as in Camus. This modern condition is at the center of Lorca's theater, particularly in *The House of Bernarda Alba,* in which terror and fossilizing structures are present, staging a complete and total power that cannot be subverted by the individual, even though his characters do not reach the terrifying gesture of Genet's.

Before going into further analysis of *The House of Bernarda Alba,* I will point out some of the functions of the maids reflected in the substructure of a couple of the plays before 1936 that announce cultural dislocation. Briefly, in *The Shoemaker's Prodigious Wife,* even though they are not maids per se, the colorful Neighbors are a set of figures who preclude the meaning of the enclosing of the affirming power structures in which the maids will later represent the presence of terror, as in Genet's plays. Indeed, the Neighbors play the roles of hypocritical and burlesque censors of the Wife, hurting her with their sharp tongues and with the sounds of their piercing voices, repeating in their words the memory of the cultural web and, specifically, in the surrealist encircling choreography of their rapid movements:

> Neighbors dressed in various violent colors and carrying large glasses of cooling drinks begin to enter through the door. They turn, run, come and go around the Wife, who is sitting, shouting. The great skirts open with their turns. All adopt a comic attitude of pain.[23]

Technically the neighbors are the disruptive element that initiates the scenic movement of the play, bringing the off-stage world in to the Wife, who does not want to open her house, does not want to leave, but wants to invent her own text, indicating a desire to distance herself from the external laws and, by implication, those of the audience. Since *The Shoemaker's Prodigious Wife* is a comic drama, the Neighbors are obviously parodying a spectacle that encircles the Wife in an asphyxiating game; the choreography of the scene and the puppet style of acting indicated by Lorca feature the texture of a carnivalized ritual.

Another example is seen in *The Love of Don Perlimplin with Belisa in the Garden* in the figure of the servant Marcolfa, who subtly plays diverse roles to control Don Perlimplin, thus indicating the impossibility of belonging to the world of the master, as well as her self-hatred and her inadequacy to open up the social structure that imprisons her. The comic structure of the play gives the impression that Marcolfa is just mimicking the language of honor, but its repetitive rituals, with both characters playing their dialogues as nonsense childish singsong, will be seen in *The House of Bernarda Alba,* but this time minus the playfulness and with an obsessive overtone.

Modern theater reflects the fragmentation of representation into framed mirrors that repeat the ritual endlessly, and ritual perpetuates the repetition of the sense of imprisonment in the case of the maids; the ritual stages hatred of the outside world, of the masters, and for each other. The confining of the players to their continuous officiating negates the possibility of action and increases the atmosphere of suffocation, since the characters only exist for the representation of that hate.

In the *The House of Bernarda Alba* the question of imprisonment has been discussed by several critics, such as Gwynne Edwards, who says:

> the house . . . is a prison and . . . Bernarda tyrannically rules her mother, her daughters and her servants. But the house is simultaneously the symbol of a condition that exists beyond it, for if Bernarda's household are her prisoners, she is in turn the prisoner of others—of the world beyond the house, of the villagers, of their gossip, their envy and their malice, and of all the values they stand for: honour, custom, tradition, all inflexibly revered.[24]

According to Edwards the metaphor of the house is that of humanity at large. He does not, however, deal with the case of Poncia as a forceful and central character in the play's struggle for power.

In a note to the reader Lorca indicates that the play purports to be a photographic documentary, an idea intensified by his use of the black and white interplay. Even more than a documentary, however, the white walls and the characters dressed in black resemble the idea of photographic negatives on top of one another where the figures in their stillness seem to move because of the continuous repetitiveness of their gesturing, only to retreat to a maddening immobility. Also, the architectonic structure of the house, with its windows and doors closed, is the frame for the madness of the characters. Bernarda keeps order while her daughters are constantly trying to phase out the walls, to reach outside of the rigid frame, and the maids keep that order in its pristine form, according to the rules of social conventions and authority.

Returning to the rituals executed by the servants, let us start with the first scene of act 1. The house is the location where the family represents itself as a family; the house, as institutional sign, depends on the actions of the maids, who are instrumental in caring for its appearance, its order and cleanliness; they are the guardians of the institution. In the first scene, Poncia, the head servant, is talking to another maid who is very busy cleaning under her supervision, for cleaning is a ritual ordered by Bernarda. The importance of cleaning is emphasized continuously. Bernarda should not find even a speck of dust or a stain that will somehow blemish the family's reputation and, by implication, the cleanliness of family honor.

Technically, in an Ibsenian style, the maids are giving information about the past to the spectator since the action starts in medias res, but this dialogue is their daily ritual to initiate the ceremonies of hate. The dialogue reflects the hate toward the social group for which they work, parodying in a burlesque way Bernarda's obsessions; Poncia's quote "she has to be the cleanest, she has to be the decentest, the highest and the mightiest"[25] is not only a burlesque comment anticipating the tragedy, but is also an incantation to begin the routine ceremony of their hateful way of living together in a relationship in which servants and masters represent their roles inflexibly.

The characters' necessary repetition of the hate incantations every day builds up the enclosure in which they are trapped into continuing to execute the ritual. It would seem that the morning ritual prepares the servants to keep representing their roles as guardians of the house for the remainder of each day, as repressed subjects who are imprisoned and powerless to rebel against the tyranny of Bernarda. However, these rituals and incantations are a sign of the madness enveloping the characters. The tone of their hatred toward Bernarda loses meaning because of its purposeless, iterative mode, since Poncia, while reciting her imprecations, is eating bread and sausage; for her the ritual is just a way to start the day and express her hate. The other maid is obsequious to Poncia and serves as her sounding board, but when she is alone, she abuses the beggar, repeating Bernarda's repressive pattern. The repetition of the hate discourse throughout the verbal structures creates a madness pervading the stage of the house. Thus the characters' own prison is constructed each day with the sound of their voices.

However, as the action of the play develops, so does the importance of Poncia, who generates the theatrical actions of the other characters which will end in tragedy. In the second act Adela, the youngest daughter, is forced to confess to Poncia her passionate love for Pepe, or her passion for life, only to be surrounded by the language of Poncia, who constructs walls around her by quoting fragments of the repressive cultural discourse: "leave your sister in peace"; "don't go against God's law"; "I have no love for any of you, but I want to live in a decent house. I do not want to be shamed in my old age!" (*CBA*, 830–31). Her discourse reveals the same obsessions as Bernarda's, as the servant repeats what she had mocked in the first act about decency and cleanliness.

Later in the act she informs Bernarda of the activities of the daughters, forcing Bernarda to react according to the role assigned her by the culture as mother and guardian of social values. Edwards has called the scenes between Bernarda and Poncia a "battle for domination and survival,"[26] and indeed Bernarda is progressively exasperated by Poncia's sarcastic comments. Their dialogues are violent and tense; their confrontations are iterative scenes that have been played over and over in the past, lacking in reason, becoming increasingly entangled in their struggle to dominate one another. However, both are representing within the limits of their roles, Bernarda as the oppressor and Poncia as the oppressed, although it is Poncia who controls the behavior of all the characters for the last third of the act.

Hate is present in all the exchanges between characters, sisters, mother and daughters, and between Bernarda and the servants, but it is Poncia who is always watching from the interstices and doors. Poncia is the one who keeps the dramatic action in constant movement, the one who initiates the attack in the scene sequences. Using the strategies of the oppressed, silence and withdrawal, Poncia is always waiting for the interstice that will let her challenge once again Bernarda's control in a ritual game that will only take them to death and, ironically, to silence. The transvestism of the surrealists' games of *The Public* is being transformed and worked into arabesquean tactics that will mirror the spectator's fears, while the repetitive scenes full of hate create a locus of cruelty in which the characters are trapped like Genet's maids, when Claire forces Solange to poison her.

Playing a dangerous game of envisioning freedom in a ritualistic way, Poncia, referring to Bernarda, declares early in the first act, "I will be spitting on her for a whole year" (*CBA*, 792). She does not see that freedom has already become hatred. However, at the end of the third act, when Poncia mimics the action of strangulation indicated in the stage directions, "putting her hands on her throat" to show how Adela died and exclaiming, "May we never die of that death" (*CBA*, 881), she knows that the only possible ritual left is that of asphyxiation, for now they are imprisoned in a labyrinth of silence where there is no salvation. The games that were seductively dangerous have now ended in silence, the ultimate repression of tyranny.[27]

If in Genet's play the lives of Claire and Solange were to signify a salvation for the repressed subject after death, it was because they had desired to be the executors of death. In the case of Poncia, the only possible signification is to keep representing rituals that are framed by the walls of a house in which desire is also trapped. In the continuous game of hatred, they have constructed similitude after similitude without progressing, each similitude being a frame, being a wall without exit, for there was no desire for freedom.

Lorca does not ask for a declamatory tone of acting as Genet does for his maids, but the gesturing, the obsessiveness of their discourses, the hallucinatory tone of their voices, and the pervading silence always surrounding them carry a consciousness of symbolic representation. The characters are not just representing "women's drama in Spanish villages" (*CBA*, 789), as the subtitle indicates, for they have become signs of madness, signs of a cultural death wish. Since games and rituals between Poncia and the other characters are played publicly, for the entire house is the stage, the representation plays in the inner mental space of the spectators, who recognize the cultural signs of asphyxiation.[28]

I have been referring to visual and metaphoric signs of strangulation and enclosure that often appear in Lorca, signaling both death and fossilizing cultural structures. Let us return to Poncia's last gesture when she was mimicking Adela's suicide with her hands. How might this strangulation gesture be reflected in painting? The concrete figures of the maids and their hate games disappear; the image of strangulation pervades the whole social text and has been transformed into pure gesture in Dalí's painting *Premonition of the Spanish Civil War, with Boiled Beans,* finished six months before the civil war.[29] The most striking characteristic of this painting is the grotesquely deformed hand strangling its own body. The cultural image of suffocation culminates in an image of cultural enclosure and suffocation in the painting *Guernica* by Picasso.[30] The images of suffocation and hatred staged in Lorca's plays announced the explosion of violence and fragmentation as an integral component of the political modern throughout Spain and Europe and the theatricality of terror in the postmodern.

Notes

1. The basis for the alignment with surrealism has been explored in my book *Lorca: El espacio de la representación* (Barcelona: Ediciones del Mall, 1986).

2. All quotations from *The Public* are from Federico García Lorca, *The Public and Play without a Title*, ed. and trans. Carlos Bauer (New York: New Directions, 1983), 11; however, I have changed the word *Character* to *Figure* as a more appropriate translation.

3. For the idea of terror as applied to modern drama, as well as its relationship to the political, see Anthony Kubiak, *Stages of Terror* (Bloomington: Indiana University Press, 1991), 120–50.

4. Oreste F. Pucciani, "Tragedy and the Maids," in *Genet/Ionesco: The Theatre of the Double*, ed. Kelly Morris (New York: Bantam Books, 1969), 23–25.

5. Mildred Adams, *García Lorca: Playwright and Poet* (New York: Brasiller, 1977), 196.

6. For theater seen as a system of performing cultural frames, see Victor Turner, *From Ritual to Theatre: The Human Seriousness of Play* (New York: *Performing Arts Journal*, 1982), 20–60.

7. José Antonio Maravall, *Culture of the Baroque: Analysis of a Historical Structure* (Minneapolis: University of Minnesota Press, 1986), 19–126.

8. Victor Turner, "Frame, Flow and Reflection: Ritual and Drama as Public Liminality," in *Performance in Postmodern Culture*, ed. Michel Benamou and Charles Caramello (Madison, Wis.: Coda Press, 1977), 33.

9. In general I follow Michel Foucault's analysis on the performance shown in the painting; see *The Order of Things: An Archaelogy of the Human Sciences* (New York: Pantheon Books, 1970), 3–16 and 308. However, I add Maribárbola as an important piece of the representation.

10. Kenneth Clarke, *Looking at Pictures* (Boston: Beacon Press, 1969), 40.

11. Miriam Balboa Echeverría, "*Casa con dos puertas*: Women, Capes and Marriage," in *The Calderonian Stage: Body and Soul*, ed. Manuel Delgado Morales (Lewisburg, Penna.: Bucknell University Press, 1997), 165–84.

12. See Kenneth Muir and L. Mackenzie, *Three Comedies by Pedro Calderón de la Barca* (Lexington: The University Press of Kentucky, 1985), 22.

13. In my analysis of and references to culture and knowledge, I follow, among others, Roland Barthes, *The Pleasure of the Text* (New York: Hill & Wang, 1975); Jean Baudrillard, *Simulacra and Simulation* (Ann Arbor: University of Michigan Press, 1999), and "Our Theatre of Cruelty," *Semiotext(e)* 4 (1982): 108–9; Walter Benjamin, "Critique of Violence," *Reflections* (New York: Schoken Books, 1986), 297; Louis Althusser, *Lenin and Philosophy, and Other Essays* (New York: Monthly Review, 1977); Fredric Jameson, *The Political Unconscious: Narrative as a Socially Symbolic Act* (Ithaca, N.Y.: Cornell University Press, 1981).

14. Turner, "Frame, Flow and Reflection," 33.

15. Fredric Jameson, *Fables of Aggression* (Berkeley: University of California Press, 1979), 13.

16. The quotations from the play are from Jean Genet, *The Maids and Deathwatch*, trans. Bernard Frechtman (New York: Grove Press, 1982), 47. Hereafter *M*, cited in the text.

17. See Silvio Gaggi, *Modern/Postmodern: A Study in Twentieth-Century Arts and Ideas* (Philadelphia: University of Pennsylvania Press, 1989), 41–45.

18. See Martin Esslin, *The Theatre of the Absurd* (Garden City, N.Y.: Doubleday, 1969), 177; Jean Paul Sartre, introduction to Genet, *The Maids and Deathwatch*, 7–8.

19. Jean Genet, *Les Bonnes* (Paris: Gallimard, 1949), 17, my translation.

20. Lucien Goldmann, "The Theatre of Genet: A Sociological Study," in *Genet/Ionesco*, ed. Morris, 106.

21. See Jean Genet, "A Note on Theatre," ibid., 19–22. In this article Genet makes some points on the declamatory and exaggerated style he wanted for the maids.

22. For the concept of simulacra see Baudrillard, *Simulacra and Simulation*, 61.

23. Federico García Lorca, *Five Plays*, trans. James Graham-Lujan and Richard L. O'Connell (New York: New Directions, 1963), 79.

24. Gwynne Edwards, *The Theatre beneath the Sand* (London: Boyars, 1980), 242. For the question of spaces see Francisco Ruiz Ramón, "Espacios dramáticos en *La casa de Bernarda Alba*," *Gestos* 1 (1986):80–82. For his part, C. Christopher Soufas has discussed the question of authority in the play in *Audience and Authority in the Modernist Theater of Federico García Lorca* (Tuscaloosa: University of Alabama Press, 1996).

25. Federico García Lorca, *La casa de Bernarda Alba, Obras completas* (Madrid: Aguilar, 1974), 791. Hereafter *CBA*, cited in the text. All quotations are my own translations.

26. Edwards, *Theatre beneath the Sand*, 260.

27. For silence and closure of representation, see Jacques Derrida,"The Theatre of Cruelty and the Closure of Representation," in *Writing and Difference* (Chicago: University of Chicago Press, 1978), 243–44; see also Antonin Artaud, *The Theatre and Its Double* (New York: Grove Press, 1958), 233.

28. When using photography images and terms, I follow Roland Barthes, *Camera Lucida*, trans. Richard Howard (New York: Hill & Wang, 1981). I do not intrepret Lorca's term "photograpic documentary" as has been interpreted by many critics who talk about photographic realism when analyzing *The House of Bernarda Alba*. This point is taken and discussed at length by Soufas, *Audience and Authority*, 170 n.3.

29. Dalí's painting is on exhibit at The Princess Sophia Museum in Madrid.

30. Picasso's *Guernica* is on exhibit at The Princess Sophia Museum in Madrid.

Embroiderers of Freedom and Desire in Lorca's Poetry and Theater

Manuel Delgado Morales
Bucknell University

ONE of the most frequent portrayals of female characters in the literary creation of Federico García Lorca is that of women engaged in embroidery, lacemaking, and sewing. Lorca's fascination with women's needlework was shared by his friends Salvador Dalí and Luis Buñuel, a fact that suggests that all three of these avant-garde artists sought to reconsider the symbolic potential of these artforms normally carried out by women. It seems likely that Lorca, as well as Dalí and Buñuel, first took an interest in the artistic portrayal of women engaged in lacemaking, sewing, and embroidery because of the revolutionary implications of this domestic activity. Their sense of discovery as young critics of art soon became transformed into a creative vision manifested in poetry and theater (Lorca), painting (Dalí), and film (Buñuel).

An example of this can be seen in the fascination of the three friends with Jan Vermeer's *Lacemaker.* As Agustín Sánchez Vidal has pointed out with regard to Dalí and Buñuel,[1] it is very probable that the common interest of the three friends in the figure of the embroiderer started at the *Residencia de Estudiantes* [Students' Residence] of Madrid, where they all lived for a time. In the *Residencia,* according to Sánchez Vidal, hung a reproduction of Vermeer's *Lacemaker,*[2] a painting that likely called the attention and piqued the curiosity of Lorca, Dalí, and Buñuel, and that would become a subject of their conversations and artistic endeavors. Keeping in mind Lorca's influence on Dalí and Buñuel, I tend to believe that the Andalusian poet was the one who awakened the painter's and the filmmaker's particular interests in Vermeer's *Lacemaker* and in the theme of the embroiderer in general. Lorca is the first to refer to Vermeer and his work, when in *Impresiones y paisajes* (1918) he compares the light of some Flemish interiors in the Covarrubias church to the light created by the *Lacemaker's* painter.[3]

For his part, Dalí declared his attraction for Vermeer's *Lacemaker* at a lecture given at the University of the Sorbonne on 17 December 1955:

> Lo que más me conmovió de ese cuadro fue que todo convergía exactamente en una aguja, en un alfiler, que no estaba pintado, sino sólo sugerido. Con mucha frecuencia sentía la acuidad de aquel alfiler con tanta realidad en mi propia carne, en mi codo, que me despertaba sobresaltado en medio de las siestas más paradisíacas. Aquel cuadro había sido considerado siempre como muy apacible y excesivamente sosegado. Para mí poseía una de las fuerzas más violentas en el dominio de la estética, a la que únicamente el antiprotón que acababa de descubrirse, puede comparársele.[4]

> [What moved me most about that painting was how everything converged exactly into one needle, in a pin that was not painted, but merely suggested. Very often I felt the pin's sharpness pierce my own flesh so realistically, on my elbow, that I would awaken frightened from the most heavenly naps. That painting had always been considered pleasant, and excessively restful. For me, it possessed one of the most violent forces in the aesthetics domain, which can only be comparable to the recently discovered antiproton.]

Dalí's interest in Vermeer's *Lacemaker* and in women engaged in needlework manifests itself not only in his trip to the Louvre to gaze at the canvas of the *Lacemaker*, but also in the numerous paintings he rendered in the 1920s portraying women embroidering and sewing. Among these paintings are *Portrait of Grandmother Ana Sewing* (1920), *Girls in a Garden* (1921), *Ana María Sewing* (1926), *Girl Sewing* (1926), *Woman at the Window at Figueras* (1926), his three studies of *Girl Sewing* (1925, 1926), as well as other paintings in which needlework acquires special relevance (*Still Life*, 1924 and *Study for the Great Masturbator*, 1929).[5] Years later, in the 1950s, Dalí would return to the theme of Vermeer's *Lacemaker* with renewed vigor, to create around her what the painter himself and Robert Descharnes came to call *The Amazing Adventure of the Lacemaker and the Rhinoceros*.[6] Among the best known of Dalí's paintings of this time are copies of the *Lacemaker* by Vermeer (1955), *Rhinocerotic Bust of Vermeer's Lacemaker* (1955), *Virgin Auto-Sodomized by Her Own Chastity* (1954), and *Rhinocerotic Portrait of Vermeer's Lacemaker* (1955).

Like Dalí and Lorca, Buñuel also showed great interest in the female characters engaged in needlework. He not only included them in his joint film venture with Dalí, *Un Chien andalou* (1929), but also in many of his subsequent films, *Woman without Love* (1951), *The*

Brute (1953), *He* (1953), *Abysses of Passion* (1954), *Death in the Garden* (1956), *The Young Girl* (1960), *Viridiana* (1964), *Diary of a Waitress* (1964), *Tristana* (1970), and *That Dark Object of Desire* (1977). Vermeer's lacemaker, indeed, is featured in *Un Chien andalou.*

With time, the figure of a woman engaged in embroidery or sewing would become a symbolic presence in the works of the three friends, as did other motifs, such as the severed hand, the cloud, the eyes, the dog, Saint Sebastian, the theme of the Holy Sacrament, or the Last Supper. The richness and complexity of this symbol was to increase significantly under the influence of Freudian and surrealist ideas.

For the present study I have chosen to focus on the works of Lorca, taking as their visual counterpart some of Dalí's paintings mentioned above. The presence of the embroiderer and her work can be observed in Lorca's short narrative *Historia de este Gallo,* as well as in his plays *Tragicomedia de don Cristóbal y la señá Rosita, Mariana Pineda, Yerma, Bodas de sangre,* and *La casa de Bernarda Alba,* and in poems such as "Amparo," "Las seis cuerdas," Crótalo," "La casada infiel," and "La monja gitana."

In my opinion, the frequent reference to needlework in Lorca's theater and poetry represents more than a simple wink to Dalí and Buñuel. It takes on feminine connotations or allusions that contribute to our understanding of the range, complexity, and richness of his work. In fact, not only do Vermeer's lacemaker or Salvador Dalí's *noias* [girls] march past Lorca's poetic retina, but also the real women of Granada, among whom we must count the mother of our poet, Doña Vicenta Lorca. In those years these women devoted themselves to needlework in order to contribute in some way to the domestic economy or in order to break the monotony and the dead hours of their daily routines.

After reading Lorca's poetic and theatrical works, it is easy to imagine him as a young boy or an adolescent among these women, attentively observing their bobbin laces, needles, drums, sewing kits, and threads, and listening to their gossip, sighs, frustrations, and desires. Needles, threads, and laces that in the poetic realm created by Lorca will be transformed into symbols (=voices) of desire and sexual liberation. Needles and threads that are reminiscent of the needles and drawers described by Freud or painted by Dalí. With these instruments the poet (=needleworker) naturally and artfully will elaborate a text (=fabric) that is very similar to that described by Roland Barthes:

> Text means Tissue; but whereas hitherto we have always taken this tissue as a product, a ready-made veil behind which lies, more or less hidden,

> meaning (truth), we are now emphasizing, in the tissue, the generative idea that the text is made, is worked out in perpetual interweaving; lost in this tissue—this texture—the subject unmakes himself, like a spider dissolving in the constructive secretions of [her] web. Were we fond of neologisms, we might define the theory of the text as hyphology (hyphos is the tissue and the spider's web).[7]

However, far from limiting myself exclusively to the study of this text, woven or embroidered, as Barthes and other poststructuralists usually do, I will also center my attention on the numerous embroiderers created by Lorca, and propose to consider them as the spider (=Arachne) studied by Sandra Gilbert, Susan Gubar, and Nancy Miller.[8] For obvious reasons, I will not attempt to find an individualized and concrete female "I" in these embroiderers, but rather the different masks or feminine poses which Lorca's poetic "I" uses to manifest or hide its own identity, desires, aspirations, and frustrations. In light of this self-referentiality or self-reflection, it can be argued that Lorca attributes to himself the feminine features inherent in the spider or the embroider, an act that has been described by Miller as the "masculine recuperation of the feminine" and by Alice Jardine as "gynesis" or "the putting into discourse of 'woman.' "[9]

One of Lorca's literary creations that best illustrates this correspondence between the needleworker's or spider's (=Arachne's) works and Lorca's writing can be found in *Historia de este gallo,* a story written by Lorca in the first issue of his journal *Gallo* to make more explicit his intellectual, literary, and artistic objectives. The name given by Lorca to the protagonist of this story is Don Alhambro, a masculine version in Spanish of the Alhambra of Granada and a character that according to García-Posada stands for Lorca himself.[10] As Lorca was trying to achieve in real life with the creation of his journal—and to a larger extent with most of his literary works—Don Alhambro intends to awaken the city of Granada "del letargo en que estaba sumergida" [from the lethargy in which it was immersed] (3:298). With that purpose in mind, Don Alhambro (=Lorca) decides to create a journal that has a rooster as its symbol. After a long search and after finding the best available rooster in Granada, Don Alhambro has to decide which, among the two embroidery schools of the city, is the most suitable to embroider the image of the rooster for this new journal:

> En aquella época venturosa, Granada estaba dividida por dos grandes escuelas de bordado. De una parte, las monjas del Beaterio de Santo Domingo. De otra, la eminente Paquita Raya. Las monjas de Santo Domingo conservaban en una caja de terciopelo las dos agujas matrices de

su escuela barroca, las dos agujas con que hicieron maravillas virginales las artistas Sor Sacramento del Oro y Sor Visitación de la Plata. Era aquella caja como el fuego vestal que inflamaba el corazón almidonado de las novicias. Elixir permanente de hilo y consulta. Paquita Raya, en cambio, tenía un arte más popular, más vibrante, un arte republicano, lleno de sandías abiertas y de manzanas endurecidas sobre el tejido. Arte de exactas realidades y emoción española. Había una gran lucha entre las dos escuelas. Todas las personas morenas eran partidarias de Paquita. Todas las rubias, castañas y un pequeño núcleo de albinas, partidarias de las monjas. Aunque hay que confesar que las dos escuelas eran maravillosas, porque si las religiosas del Beaterio triunfaban empleando una tonelada de oro en el manto de la Soledad de Osuna, Paquita triunfaba en Bruselas con un bordado representando el Patio de los Leones, en el cual había más de cinco millones y medio de puntadas. (*OC,* 3:300–301)

[In that fortunate time, Granada was divided between two great embroidering schools. On the one hand, the nuns of Santo Domingo Convent. On the other, the great Paquita Raya. The nuns of Santo Domingo kept in a velvet box the two chief needles of their baroque school, the two needles with which the artists Sister Sacrament of Gold and Sister Visitation of Silver made virginal wonders. That box was like the vestal fire that inflamed the starched heart of the novices. It was a permanent elixir of thread and consultation. Paquita Raya, on the other hand, had a more popular and vibrant art, a republican art full of open watermelons and hardened apples on the fabric. Hers was an art of exact realities and Spanish emotion. There was a great struggle between the two schools. All the dark-haired people favored Paquita. All the blond and light-brown-haired people, as well as a small nucleus of albinos, favored the nuns. However it must be said that the two schools were wonderful because if the nuns of the Convent triumphed by using a ton of gold on the cape of our Lady of Osuna, Paquita triumphed in Brussels with an embroidery that represented the Court of the Lions, which had more than five and a half million stitches.]

As this passage demonstrates, Don Alhambro's preference for Paquita Raya's work is due to the fact that hers is a more republican, popular, and socialist art, while the nuns' is, in turn, more religious and in consonance with traditional Spanish values. Paquita's embroidery, and, by the same token, Don Alhambro's and Lorca's literary endeavors, represent a defiance of the Spanish and Granadian bourgeoisie, a defiance that can be considered as a reenactment of Arachne's challenge to Athena's art and authority.

One of the most prominent features of Lorca's theatrical and poetic needleworkers is that they all do their work in a small, closed, suffocating place which can be considered a direct consequence of social, religious, political, or psychological repression. In Mariana Pineda, for example, the Granadian heroine who gives the play its name, sews and embroiders the liberty flag, hidden in the most remote room of her house. As the play shows, Mariana's house is located on a "silent" street, in one of the most isolated neighborhoods of one of Spain's most labyrinthine cities, Granada. According to Angustias, the only way to see the flag is "por el ojo de la llave" [through the keyhole] (*OC*, 2:86). The confined, asphyxiating atmosphere of Mariana Pineda's house and the city of Granada is accentuated through the diegetic and mimetic representations of Santa María Egipciaca convent and the Santa Catalina tower, where Mariana and Pedro de Sotomayor are held prisoners. It must be noted, however, that Lorca alters certain historical facts to accentuate Mariana's tragic condition. For example, instead of having other women embroider the flag, as the historical version summarized by Andrés Soria Olmedo shows,[11] Mariana does the work with her own hands, seeing herself obligated to do so for her liberal friends and, above all, for the man that she loves blindly, Pedro de Sotomayor. Likewise, Mariana is not inspired to embroider the flag for the political reasons that led Pedro de Sotomayor and the rest of the conspirators to rebel against the king. Rather, she embroiders the flag to be free to love whomever and however she pleases as a young widow.

One could say, on the other hand, that upon seeing herself oppressed between two contrary forces—the conservative tradition and the liberal uprising—Mariana has to work within the established order to subvert it. In the end, she is a victim of the repressors' cruelty and of the liberals' obscurity, ingratitude, and impotence. Curiously, the river Genil is the only escape in Granada from the sociopolitical labyrinth that surrounds Mariana and her friends. Although Pedro de Sotomayor escapes with Mariana's help, she will never get to the river, because she has been imprisoned by Pedrosa.

In addition to the historical context, in *Mariana Pineda* one perceives a mythical structure. The heroine from Granada takes the place of Ariadne with her helping Pedro de Sotomayor so that he would find freedom. At the same time, she expects this new Theseus to return the love she demonstrated to him. Nevertheless, Mariana is abandoned at Naxos, as was Ariadne, by the man whom she helped to escape, the difference being that if the Greek Ariadne let herself be carried away or loved by Dionysus, the Granadian Mariana is incapable of hearing Don Fernando's loving words, promising freedom and the fulfillment of the love that Don Pedro could not give to her.

Without matching the tragic tone of *Mariana Pineda*, the poem

"Amparo" describes its protagonist embroidering letters on a canvas. Now, what immediately captures the spectator-poet's attention is the loneliness and whiteness around Amparo, who is described as "ecuador entre el jazmín y el nardo" [equator between jasmine and spikenard] (*OC,* 1:326). The loneliness, neglect, and silence that accompany Amparo's virginal whiteness are only interrupted by murmurs of the fountains in the house's courtyard and by the canary's yellow trilling. The yellow color is made present in the poem through what Carlos Bousoño has described as "desplazamiento calificativo" [qualifying displacement].[12] The aforementioned color can make an allusion, on one hand, to the effect of the passage of time in Amparo's youth. On the other hand, it could suggest the jealousy and bitterness that consume the embroiderer when she finds herself abandoned by the indecisive man to whom José María Guarnido alludes.[13] Actually, the space in which Amparo is embroidering not only is defined by the young woman's white dress or by the embroidering canvas on which she completes her work; it is also defined by her own demeanor, which Lorca describes as "equator" between two poles of white, that is to say, between jasmine and spikenard. Keeping in mind the possible significance of the white color in works such as "La monja gitana" and *La casa de Bernarda Alba,* Lorca's portrayal of Amparo as an "equator" between the two poles of white can be interpreted as a mode of being between her own virginity and a kind of death, symbolized by the whiteness in the house of Bernarda Alba.

One could say, then, that Amparo's embroidery represents at the same time the dream of a love or desire that will never be realized and the frustration and resignation of the young woman at this deferment. For this reason, Amparo can be compared to Doña Rosita the Spinster and to one of Lorca's drawings, *Muchacha granadina en un jardín.*[14] I also believe that her virginity and seclusion is reiterated twenty years later in Dalí's *Young Virgin Auto-Sodomized by Her Own Chastity* (1954) as well as in the paintings in which Dalí relates the feminine forms of Vermeer's lacemaker to the horns of the rhinoceros.

Amparo's loneliness, isolation, virginity, and repression are more developed in the poem "La monja gitana," whose title seems to announce at once the strong oxymoron on which Cecelia Cavanaugh comments.[15] It must be made clear, however, that neither the constitutions nor the unspoken law of female religious orders envisaged that a Gypsy would become a nun, just as they never considered that a nun could be a Gypsy. In other words, saying "Gypsy nun" is equivalent to juxtaposing two mutually exclusive words. In this way, more than a name given to people from a particular ethnic group, the word *Gypsy* must be considered a qualifying adjective that describes the mindset and behavior of a woman who, feeling obligated to repress her desire in an extremely small and suffocating space, tries to overcome her repression by fulfilling her desire.

The spatial and personal whiteness that defines and limits Amparo is converted in "La monja gitana" into "silencio de cal y mirto" [silence of lime and myrtle] (*OC*, 1:423), words that offer the reader various connotations and interpretations. In the first place, upon pronouncing or hearing "cal y . . ." [lime and . . .] in Spanish, both the speaker (=poet) and the listener immediately think of the familiar expression "cal y canto" [lime and stone], which is used to speak of something strong, solid, and very durable. In fact, when Lorca replaces the stone with the myrtle, in order to define the silence and space surrounding the embroidering nun, he creates a double wall around her. The first wall is made of masonry and the second of myrtle, which can reach the height of two or three meters on the walls and fences of Granada. The wall that surrounds the Gypsy nun has a more marked content of lime than that in the space where Amparo embroiders, and thus it brings to mind the thick, white wall of Bernarda Alba's house and the caustic lime applied to Ignacio Sánchez Mejías's body. In both cases, the lime not only indicates virginity or cleanliness, by virtue of its whiteness, but also sterility, death, and a cemetery.

After describing the double wall that separates the nun from the outside world, Lorca describes the space closest to her, a space which is composed of "hierbas finas" [fine herbs] (*OC*, 1:423) and mallow. Although this space can, on the one hand, correspond to a convent's garden, it reflects, on the other hand, the mental condition of the embroidering woman. Her vital impulse and desire, the fine herbs, are speckled with mallows, a color that also defines Mariana Pineda's state of mind. In any case, as María Moliner explains, the mallow color applies to someone docile and submissive, while the phrase "criar malvas" [to raise mallows] can refer to someone who is dead or buried.[16]

Reduced to such a mental and physical state, the nun dedicates herself to embroidering tiny wallflowers on the altar cloth. As we have already seen, Mariana Pineda embroiders to overthrow the king's authority and to carve out her own freedom from the labyrinthine and suffocating Granada. Likewise, the nun embroiders as she dreams of changing the cloth's whiteness into the colors of the rainbow, trying to free herself from the mental and moral prison, formed by the many convent walls. In other words, inspired by the colors produced by the refraction of the light in the window's prism or from the "araña" [chandelier][17] hanging from the roof, the nun becomes a Gypsy in the exact moment that she tries to break away from the walls and from the convent restrictions. She becomes a Gypsy because she dreams of embroidering sunflowers, magnolias, saffron, ribbons, and moons on the mass cloth, just as the church, "como un oso panza arriba" [like a bear on its back] (*OC,* 1:423), reminds her of her moral obligations. Moreover, she becomes a Gypsy because as soon as she must choose between the sensory pleasure of the kitchen—the five grapefruits from Almería sweetened in the kitchen—and the spiritual sacrifice of the mass, she opts for the former, desecrating or rejecting the latter. It is in this sense that one must interpret the juxtaposition of such different spaces as the kitchen, the church, and the place where the nun is embroidering. As Bousoño has indicated, in our time this spatial juxtaposition results from "su gusto por la sugerencia: el ámbito, en cuanto correlata con el protagonista lírico . . . nos dice, implícita y borrosamente, 'irracionalmente,' algo de él" [the taste for suggestion; the space, as it relates to the lyrical protagonist . . . tells us, implicitly, vaguely and irrationally, something about him].[18]

Accentuating the contrast between the space of the church and that of the kitchen, or between the repression of desire and the need to satisfy it, Lorca insists on the force of the latter when saying of the nun that a final and dull rumor lifts her blouse. It is the moment at which the nun contemplates the outer landscape, formed of clouds, mountains, and "dead distances," and when she imagines the plain raised having twenty suns and standing rivers.

This constant movement between the repression and the need to satisfy her desire or her sexuality, between the oppressive reality of the convent and the psychological need to escape it, causes the outer world and the inner one to become one in the eyes of the nun, when two horseback riders are reflected in her eyes or when, thanks to the superimposition of levels, her own eyes become two horseback riders that try to overcome any distance or obstacle. It is at this moment when she becomes a Gypsy, that is to say, when she aspires, as Mariana Pineda, to the freedom of love or the fulfillment of desire that can be given to her by a horseback rider like Pedro de Sotomayor, and also when it recreates before her eyes the arena and the bullfight of Ronda. It is then when our nun profanes, or "agitana" [makes Gypsylike], the tablecloth of the mass, according to what has been observed by Allan Josephs and Juan Caballero.[19] However, like Mariana Pineda, Amparo, and Yerma, the Gypsy nun is forced to remain locked up within her space and circumstance, without being able to cross a river and, what is worse, without being able to see it. With the double flavor of the sugar and "yerbaluisa" [lemon verbena] in her mouth, this Gypsy nun, the equivalent to Dalí's *Auto-Sodomized Young Virgin* and to his *Woman at the Window at Figueras,* tries to burn her sewing kit as does, years later, the sleepwalker, Viridiana, of Buñuel. But for her repression and frustration, she is forced to continue embroidering the tablecloth of the mass in front of the lattice window. This lattice window becomes a table or a chessboard with the effect of the afternoon light, the same light that in the beginning of the poem had been refracted in multiple colors in the prism of the "araña" [chandelier].[20]

One of the best-known examples of Lorca's embroiderers who never transcends the space that limits and ties her down is the female protagonist of *Yerma.* As can be seen in this Lorca tragedy, at the beginning of the play Yerma displays a very similar attitude to that of the Gypsy nun at the end of the poem. In other words, Yerma appears dreaming in an atmosphere of "una extraña luz de sueño" [a strange dreamlike light] and "con un tabaque de costura a los pies" [with a sewing kit at her feet] (*OC,* 2:480). The fantasies of the Gypsy nun, in other words the "Gypsy" flowers that she had wanted to embroider on in the tablecloth of the mass and the horseback riders that march by her eyes, are both replaced in the play *Yerma* by the dreams of the protagonist of the child and Víctor, that it is to say with the man who seemed most suitable to satisfy her desire, since he made her tremble, when, as teenagers, he helped her to cross the river.

Once again, Lorca establishes a close relationship between the needlework of Yerma and her dreams to fulfill her desire. In this way, when something or somebody interrupts her dream, Yerma leaves the needlework momentarily and returns to it and to her dream as soon as they leave her alone. As we have seen, when the curtain rises in the first act, "está Yerma dormida con un tabaque de costura a los pies" [Yerma is sleeping with a sewing kit at her feet]. The scene, continues the stage direction, "tiene una extraña luz de sueño. Un pastor sale de puntillas, mirando fijamente a Yerma" [has a strange dreamlike light. A shepherd enters tiptoeing and looking at Yerma] (*OC,* 2:480). However, when the clock sounds, the shepherd disappears and Yerma wakes up Juan so that he may go to work in the field. When Juan leaves the house, Yerma again returns to her sewing, at the same time that she begins to daydream, with un "hermoso bostezo" [a beautiful yawn], and "se pasa la mano por el vientre" [runs her hand over her abdomen] (*OC,* 2:482).

Interrupted by María, who comes to tell her that she just has bought "encajes, tres varas de hilo, cintas y lana de color para hacer madroños" [lace, three thread twigs, ribbons and colored wool to make strawberries] (*OC,* 2:483) because a son has arrived for her, Yerma remains "en la misma actitud que al principio" [in the same attitude as in the beginning] (*OC,* 2:486) of the act, that is to say, when María goes away, cutting diapers, and surrounding them with embroidery. After María leaves, Víctor enters the stage and once he also departs, Yerma "coge otra vez la costura. Comienza a coser y queda con los ojos fijos en punto" [takes again the sewing. She begins to sew and remains with her eyes fixed on a point] (*OC,* 2:486).

As we can see, Yerma is obsessed with sewing, as much as she is obsessed with having a son, being fertile, ceasing to be "yerma" [barren], and being able to satisfy her desire. It can be said, then, that her embroidery represents her aspiration for the fertility and the fulfillment enjoyed by María, who has bought "embroidery, three thread twigs, ribbons and colored wool to make strawberries" in order to celebrate her fecundity. Yerma, on the contrary, must begin to embroider time and time again, without carrying out a concrete design, a fact that reflects her state of permanent frustration, her sterility, and her existential emptiness. In other words, Yerma will never have in her hands the live bird of which María speaks, much like the different embroiderers we are analyzing. For that reason, when Yerma describes her barren condition and her frustration, she does it with the symbols of water and the thread to embroider:

> Quiero beber agua y no hay vaso ni agua; quiero subir al monte y no tengo pies; quiero bordar mis enaguas y no encuentro los hilos. (*OC,* 2:504)
>
> [I want to drink water and there is neither glass nor water; I want to climb the mountain and I do not have feet; I want to embroider my underskirt and cannot find the threads.]

Yerma is unable to fulfill her desires because she lacks the necessary foundations or potentiating elements. In one way or another, she is anxiously looking for a needle, the same invisible needle, "not painted, but merely suggested," that Dalí claims to have felt deeply when contemplating the painting of the *Lacemaker*—the needle that Freud would identify with the phallus[21]—the same needle that Dalí frequently related to the rhinocerotic horns, which in turn constituted and defined his *Virgin Auto-Sodomized* and the numerous "copies" that he made of Vermeer's *Lacemaker.* Needle, desired and feared, at the same time, in *Yerma,* when Víctor tells the protagonist that Juan should go deeper ["que ahonde"] (*OC,* 2:486) if she wants to have a child, and when, preannouncing the final tragedy, Yerma makes the *Muchacha* 1 [Young Girl 1] notice that a simple "agujita" [little needle] can be the cause of the death of her son. Needle, that because it has been continuously postponed, forgotten, or sublimated ends up by becoming "juncos grises del invierno" [gray rushes of winter], the "needles" put by the oaks underneath the pillow of the shepherd, and, finally, the bitter "espina de la retama" [thorn (= needle) of broom] (*OC,* 2:492) that causes the death of the shepherd.

Robert Lima is correct when he affirms that two opposing forces or elements are in conflict in Yerma, the Christian ethos and the pagan tradition of Dionysus.[22] In my opinion, this is the dilemma faced, in one way or another, by the different embroiderers created by Lorca. Except for the "casada infiel" [faithless wife] or "the embroidered" *Novia* [Fiancée] of *Bodas de sangre,* few of these embroiderers solve the dilemma in favor of their desire or of the impulses of nature. For one reason or another, Mariana, Amparo, the Gypsy nun, and Yerma are trapped in their own labyrinth or prison—convent, courtyard, psyche, or idiosyncrasy—without managing to arrive at the arena of Ronda,[23] at the river or the wall where the son of the *Vieja* [Old Lady] is awaiting Yerma. A similar fate awaits Doña Rosita, "la soltera" [the spinster], of whom Lorca says:

> Doña Rosita es la vida mansa por fuera y resquemada por dentro de una doncella granadina . . . el drama de la cursilería española, del ansia de gozar que las mujeres han de reprimir por fuerza en lo más hondo de su entraña enfebrecida. (*OC,* 3:620–21)
>
> [Doña Rosita is meek and mild life on the outside and burnt on the inside of a virgin girl from Granada . . . the drama of Spanish affectation, the eagerness for enjoyment that women must necessarily repress in the depth of their feverish womb.]

According to Lima's observation with regard to Yerma, all these embroiderers go against nature when repressing their instinct or their desire.[24] Consequently, they are like "trossos de quòniam," a term used by Lorca, Dalí, and Dalí's sister, Ana María, to describe "people not in command of their faculties, who lead a physical, indeed vegetable existence."[25] Descharnes and Néret place several paintings made by Dalí in 1925 and 1926 within the group of these "trossos de quòniam," among them: *Figure on the Rocks, Women Lying on the Beach, Woman in a Chemise, Lying, The Girl of the Ampurdán,* and *Sheets of Studies,* in which the front and back of a young woman are superimposed on the same figure.[26] If one observes closely, in the latter two paintings, Dalí represents the body of the young woman using the forms of particular fruits—pears, oranges, grapefruits—something that he encountered surely in his readings of Freud and that remind us of the quinces and grapefruits Lorca assigned to Mariana Pineda and to the Gypsy nun.

As Lima has pointed out, Lorca regrets, throughout *Yerma,* that these embroiderers are not able "to feel the lessons of the ancients" (*OC,* 2:509), or, in other words, that they are not able to allow themselves to be taken by the spirit of Dionysus.[27] Lima's observation can be better understood if one takes into account Magdalena's words in *La casa de Bernarda Alba.* Made desperate by the repression created by her mother, Bernarda, Magdalena refuses to embroider, especially after having walked through all the bedrooms of the house and having discovered an embroidered tow-cloth of her grandmother. According to Magdalena, on this tow-cloth are the embroidered figures of a black man fighting with a lion and a small long-haired dog. This embroidery leads her to dream, like Yerma, of the golden age of the ancients, an age that is very far from the present time of Bernarda, because in these days "nos pudrimos por el qué dirán" [we rot because of what people will say] (*OC,* 2:137).

As I have observed throughout this essay, the embroiderers of Lorca always strive to leave the limited, labyrinthine, or virginal space that imprisons them, be this space the city of Granada, the convent of Santa María Egipciaca, the convent of the Gypsy nun surrounded by limestone and myrtle, the inner and existential prison of Yerma, or the house of Bernarda Alba. The striving to leave or to escape these different spaces, reflects, on the one hand, the innate tendency of these women to seek to satisfy their desires, and, on the other, Lorca's fascination with movement, energy, and creativity, as Miriam Balboa Echeverría has observed in the writings of the Granadian author.[28] Following Gilles Deleuze and Félix Guattari's definition of desire, homosexuality, and femaleness, Elena Gascón Vera has also pointed out that Lorca locates this desire in the totality of woman's body and identifies it with a sexuality that is expressed in terms of interaction, transformation, and becoming.[29] Ironically, however, the movement, the creativity, and the energy mentioned by Balboa Echeverría are frequently suspended, immobilized, and paralyzed in the embroiderers of Lorca, as can be seen in the cases being considered here of Mariana Pineda, the Gypsy nun, Amparo, and Yerma, as well as those of Magdalena and Amelia of *The House of Bernarda Alba.*

In light of what we have seen, it is appropriate to make a final reflection on the possible reasons that led Lorca, Dalí, and Buñuel to focus their attention on the lacemaker of Vermeer or on women engaged in embroidery or sewing in general. Leaving aside the subject of light in the painting, which seems to be one of the most obvious causes of this fascination, the figure of the lacemaker represented, on the one hand, the idea of the traditional, humble, virginal, and submissive woman, apparently uninterested in the outside world. This retirement and submission, along with the instruments of her embroidery and her own psychosomatic constitution, could be interpreted as the origin and form of her own desire or like something destined to subvert her supposed submission, humility, passivity, and virginity. As Dalí stated in his famous speech at the Sorbonne, "that painting had always been considered very peaceful and excessively calm." Nevertheless, for him, as for Lorca and Buñuel, "it possessed one of the most violent forces in the aesthetics domain, which can only be comparable to the recently discovered antiproton."[30] In other words, and according to what Dalí tried to communicate to his admiring audience at the Sorbonne, the body of Vermeer's lacemaker constituted the best example of biological dynamism because it was formed by the most violent curves, that is to say, by those of the logarithmic spiral, which the Catalan painter would identify with the pronounced and phallic curve of the rhino.

To conclude, we can say that in the images of the needle, the thread, the drawer, in the grapefruits from Almería, in the cauliflower taken by Dalí to the lecture he gave at the Sorbonne, in the rhinocerotic horn, in the fingers of the lacemaker, is where the story of the Lorquian embroiderer begins, an embroiderer that became a guitar/spider in order to trap the floating sighs of its "negro aljibe de madera" [black wooden water tank] ("Las seis cuerdas," *OC,* 1:322). An embroiderer that also became a "crótalo" [castanet] that rippled "el aire cálido" [the warm air] but that ended up drowning in its trill of wood ("Crótalo," *OC,* 1:334).

Notes

1. Agustín Sánchez Vidal, *Buñuel, Lorca, Dalí: El enigma sin fin* (Barcelona: Planeta, 1988), 337–42.

2. Ibid., 337.

3. Federico García Lorca, *Obras completas,* 4 vols., ed. Miguel García-Posada (Barcelona: Galaxia Gutenberg/Círculo de Lectores, 1997), 4:80. Hereafter *OC,* cited in the text.

4. The text of Dalí's conference was collected in *Sí* (Ariel: Barcelona, 1976), 41–42. My quotation is from Sánchez Vidal, *Buñuel, Lorca, Dalí,* 337.

5. All references to Dali's paintings are taken from Robert Descharnes and Gilles Néret, *Salvador Dalí, 1904–1989. The Paintings,* 2 vols. (Köln: Taschen, 1994).

6. See Descharnes and Néret, *Salvador Dalí,* 476–83.

7. Roland Barthes, *The Pleasure of the Text,* trans. Richard Howard (New York: Hill & Wang, 1973), 64.

8. See Sandra Gilbert and Susan Gubar, *The Madwoman in the Attic: The Woman Writer and the Nineteenth-Century Literary Imagination* (New Haven, Conn.: Yale University Press, 1979); Nancy K. Miller, "Arachnologies: The Woman, the Text, and the Critic," *The Poetics of Gender,* ed. Nancy Miller (New York: Columbia University Press, 1986), 270–95.

9. See Miller, "Arachnologies," 271; Alice Jardine, *Gynesis: Configurations of Woman and Modernity* (Ithaca, N.Y.: Cornell University Press, 1985).

10. García-Posada, introduction to Lorca, *Obras completas,* 3:15.

11. Andrés Soria Olmedo, introduction to Federico García Lorca, *Mariana Pineda* (Madrid: Espasa Calpe, 1990), 24.

12. Carlos Bousoño, *Teoría de la expresión poética* (Madrid: Gredos, 1976), 87–93.

13. Quoted by Allen Josephs and Juan Caballero in their introduction to Federico García Lorca, *Poema del Cante Jondo. Romancero gitano* (Madrid: Cátedra, 1995), 185.

14. For a reproduction of this drawing, see Cecelia J. Cavanaugh, *Lorca's Drawings and Poems* (Lewisburg, Penna.: Bucknell University Press, 1995), 64.

15. Ibid., 122–28.

16. María Moliner, *Diccionario de uso del español,* 2 vols. (Madrid: Gredos, 1981).

17. In *Lorca's Drawings and Poems,* Cavanaugh has already pointed out that "the word *araña* recalls the image of the spider—one who weaves, Arachne of classical mythology, weaving her cloth—and infers the weaving of narration" (122). It must be also noted that Sigmund Freud, in his *General Introduction to Psychoanalysis,* trans. G. Stanley Hall (New York: Boni & Liveright, 1920), considered the "hanging lamp" as a symbolic substitute for the male member (27).

18. Bousoño, *Teoría de la expresión poética,* 112.

19. Josephs and Caballero, introduction, *Poema del Cante Jondo,* 241, n.16.

20. See note 17.

21. Freud, *Introduction to Psychoanalysis,* 126–27.

22. Robert Lima, "Towards the Dionysiac: Pagan Elements and Rites in *Yerma,*" in *Lorca's Legacy: Essays on Lorca's Life, Poetry, and Theatre,* ed. Manuel Durán and Francesca Colecchia (New York: Lang), 115–33.

23. In *Mariana Pineda,* the famous arena of Ronda is portrayed as a symbol of fulfillment and liberation of women's desire.

24. Lima, "Towards the Dionysiac," 115.

25. See Descharnes and Néret, *Salvador Dalí,* 1:104.

26. Ibid., 100–112.

27. Lima, "Towards the Dionysiac," 130.

28. Miriam Balboa Echeverría, *Lorca: El espacio de la representación* (Barcelona: Ediciones del Mal, 1986), 19.

29. Elena Gascón Vera, "Stories of Madness: The Feminine in *Poet in New York,* in Durán and Colecchia, eds., *Lorca's Legacy,* 176.

30. See Sánchez Vidal, *Buñel, Lorca, Dalí,* 337.

If the Shoe Fits: Buñuel, Almodóvar, and the Modernist/Postmodernist Question

Sidney Donnell
Lafayette College

> Buñuel was very attracted to feet, particularly women's feet. Feet and all their paraphernalia; stockings, shoes, shoe store sales ladies, etc. He also liked legs quite a lot, and stockings. For Buñuel, a woman starts at her feet.
>
> —Pedro Almodóvar, 1998

TAKEN from the first lines of an essay written by Spanish director Pedro Almodóvar promoting the North American release of his motion picture *Carne trémula* (*Live Flesh,* 1997), the above epigraph serves as an amusing reminder of Luis Buñuel's obsession with obscure objects of desire, but it also serves as an indirect means of approaching the role objects play in the ways we communicate.[1] Like Almodóvar, scholars note similar repetitions and variations of the same objects throughout Buñuel's long film career, offering numerous readings as to their possible symbolism.[2] However, these perplexing images—when, for example, Buñuel places a lady's shoe under a living room couch in the comic film noir *Ensayo de un crimen* (Rehearsal of a crime/The criminal life of Archibaldo de la Cruz, 1955)—almost never lead to discussions of the empty space such objects occupy; that is to say, critics rarely ponder why the subject (usually determined by the male gaze) simply ignores one object and finds that another is a sublime fit.[3] More importantly, those few who psychoanalyze the very emptiness of the object (in terms of how it serves as the embodiment of a certain lack in the Lacanian sense) usually fail to situate such instances within a certain historical or ideological framework, focusing instead on Buñuel's art as if it were a world of freely floating signifiers which make no reference to a specific place and time.[4] Such conclusions would allow us, perhaps even inadvertently, to characterize Buñuel as a postmodernist. But could we also think of Buñuel as a modernist, that is to say, an artist who enjoys the linearity of narrative and whose feet are firmly planted in the historical?

Certain contradictions are also produced when we ask a matching question of Almodóvar, the youngest and arguably the last in a line of *auteurist* directors from Spain that began with Buñuel. In fact, in light of recent trends in Hispanic cultural studies as regards our indebtedness to both psychoanalysis and poststructuralism, film director Pedro Almodóvar is one figure whose work critics often classify as "postmodern." However, this is done without problematizing the artificial split between modernism (for instance, linearity, narrative, historical referencing) and postmodernism (for example, nonlinearity, suspended images, self-referentiality).[5] Furthermore, given the pervasiveness with which poststructuralism has penetrated contemporary intellectual thought, it is, according to Slavoj Zizek, high time for scholars interested in history and ideology to recuperate certain thinkers whose work has been appropriated by the poststructuralist project and whose messages have been lost to the so-called indeterminacy of language.[6] This concept is also espoused in postmodernism and is expressed as the impossibility of the subject—and, therefore, the author or artist—to communicate fully what s/he intends to say. With the passage of time, however, the critical tendency to use such a lens without recognizing its deficiencies begins to distort our readings of contemporary artists, such as Pedro Almodóvar, and their relationship to earlier cultural icons, including film director Luis Buñuel. Indeed, these same film directors embrace certain aspects of the postmodern aesthetic from a modernist subject position, thereby affirming their capacity to convey an intended message.

In other words, do Buñuel and Almodóvar's cinematic worlds consist solely of objects in suspended animation, thereby allowing us to situate their movies within the realm of postmodernity? Or is there an "object" (that is, a referent) to their respective cinematic languages, permitting us to view their work as historically—and politically—invested art forms within the context of modernity? Almodóvar's invitation to compare both directors leads me to argue against an exclusively postmodernist classification for both; indeed, while Buñuel and Almodóvar often attempt to produce different aesthetic effects by continually invoking objects in their films, they are ultimately unafraid to operate within a modernist framework by gesturing to the material basis of the object in history and real events. Although Buñuel and Almodóvar explore objects of desire in order to create a cinematic vocabulary and thereby communicate that which is utterly impossible to express through conventional means, their respective films ultimately, and in varying degrees, demonstrate a modernist belief in the immanence of language, that is to say, the linguistic possibility for us to say what we mean and for others to comprehend our full meaning.

Still, Almodóvar's fascination with Buñuel seems readily apparent when we examine the psychoanalytic field and a discussion of the object-language of desire. As Zizek contends:

> a whole series of Buñuel's films . . . are built around the same central motif of the—to use Buñuel's own words,—"non-explainable impossibility of the fulfillment of a simple desire." In *L'Age d'or* the couple want to consummate their love, but they are again and again prevented by some stupid accident; in *The Criminal Life of Archibaldo de la Cruz* the hero wants to accomplish a simple murder, but all his attempts fail; in *The Exterminating Angel*, after a party, a group of rich people cannot cross the threshold and leave the house; in *The Discreet Charm of the Bourgeoisie* two couples want to dine together, but unexpected complications always prevent the accomplishment of this simple wish; and finally, in *That Obscure Object of Desire*, we have the paradox of [a] woman who, through a series of tricks, postpones again and again the final moment of reunion with her old lover. (*SO*, 194)

Desire in Buñuel's films manifests itself as an insatiable appetite, a constant craving, or a Lacanian lack; by comparison, Almodóvar demonstrates how he follows a similar historical and cultural path in his representation of desire. However, in each film Zizek cites above, Buñuel parodies the bourgeoisie and its paralysis due to its trite obsessions with the "non-explainable impossibility of the fulfilment of a simple desire." It would seem reasonable, then, that Almodóvar's invocation of Buñuel is the result of more than a parallel interest in the law of desire; indeed, Almodóvar deploys psychoanalysis in *Live Flesh* in order to make a political statement about the life and times of the characters the film portrays.

According to Peter Evans, whereas psychoanalytically informed approaches to Spanish-language cinema studies have yielded particularly stimulating results, there is also a critical tendency within this discipline to build an internal, hermetically contained environment without establishing a specific relationship to history.[7] To this end, psychoanalytical approximations must also be grounded in historical and cultural specificity in order to understand the work of artists like Buñuel whose career spanned the greater part of the twentieth century. Indeed, no less can be demanded of scholarly inquiry into the work of more contemporary movie directors like Almodóvar. For his part, Zizek's struggle to recover the work of certain intellectuals, who he considers to have been mislabeled as poststructuralists, focuses primarily on the psychoanalytic theories of Jacques Lacan. Moreover, the application of his discussion of Lacan serves us well when probing the ways in which Buñuel and Almodóvar's respective approaches to creating a cinematic language are related to objects within a modernist historical context. As Zizek contends:

> post-structuralists can calmly abandon themselves to the infinite self-interpretative play of language . . . Every "objective" statement about things includes some kind of self-distance, a rebounding of the signifier from its "literal meaning." In short, language is always saying, more or less, *something other* than what it means to say.
>
> In Lacan's teaching, however, the proposition "there is no metalanguage" is to be taken literally. It means that all language is in a way an object-language: *there is no language without object.* Even when the language is apparently caught in a web of self-referential movement, even when it is apparently speaking only about itself, there is an objective, non-signifying "reference" to this movement. (*SO*, 158)

As time goes on, divergent interpretations—or even, as Zizek claims, misreadings—of major thinkers like Lacan negatively influence our understanding of other significant modern-day artists and intellectuals. In this context, poststructuralism has had its sway over psychoanalysis and has resulted in an interpretative field in which critics run the risk of obscuring the underlying modernism of Buñuel's and, especially, Almodóvar's respective works.

Nevertheless, critics—including those invested in a poststructuralist analysis of Almodóvar—recognize that *Live Flesh* represents a significant break with the director's previous films. As Paul Julian Smith states:

> Almodóvar, proclaimed for so long as the quintessential postmodernist, seems to have rediscovered . . . modernity. While a comedy like *Women on the Verge of a Nervous Breakdown* (1988) favoured irony, fragmentation and superficiality, *Live Flesh* tends towards direct engagement with its characters, coherent integration of plot, and a sense of psychological, social and historical depth.[8]

Whereas Smith identifies a sudden shift in Almodóvar's career with the making of *Live Flesh*, it is also true that Almodóvar has followed an uneven trajectory during his almost two decades as a director. Indeed, if the director has "rediscovered modernity," then Smith implies that Almodóvar's "postmodernist" period is both preceded and followed by more modernist periods. As Smith claims, the change manifests itself, in part, as "a sense of psychological, social and historical depth." By "depth," Smith appears to mean an engagement in storytelling and history as regards the content of Almodóvar's recent film; by "superficiality," he seems to refer to the filmic style and aesthetic vision of previous works as outweighing the content. Furthermore, Smith also proposes an alignment of "depth" with "modernity" and "superficiality" with "postmodernity." In short, after years of relying more on form than content, Almodóvar's newfound depth has tipped the scale in favor of modernity.

Although the course of his filmmaking has taken different turns for a variety of reasons, Almodóvar has maintained certain characteristics in common with that of Spanish surrealists like Buñuel who preceded him by two generations. For example, his low budget, overtly political, class-conscious comedies—especially *¿Qué he hecho para merecer esto?* (What have I done to deserve this?, 1985)—are very Buñuelian in terms of their use of dark humor and the chance appearance of objects. These intriguing movies yield to relatively high budget, bourgeois comedies like *Mujeres al borde de un ataque de nervios* (Women on the verge of a nervous breakdown, 1988) that were very successful with critics and at the box office. However, in films like *Tacones lejanos* (High heels, 1991) the director experienced a loss of critical acclaim in the 1990s after embarking on more melodramatic productions that merge the pathos of highly emotional (and often overwrought) storytelling with a postmodern aesthetic. Nevertheless, as Smith argues, "With his new film 'Live Flesh' Pedro Almodóvar shows he can fuse visual and sexual anarchy with the most elegant of plots."[9] In fact, Almodóvar appears consciously to break the postmodernist mold that has been ascribed to him, siding with history as a way of interpreting his own filmic language. More specifically, in *Live Flesh* the filmmaker returns to narrating the harder truths of his earlier career by looking at how Spain has matured politically and sexually since the end of the Franco regime. Perhaps not so coincidentally, it is precisely in this groundbreaking film that Almodóvar chooses to pay homage to the old master by visually quoting Buñuel's *Archibaldo de la Cruz.*[10]

Soon after the opening credits of *Live Flesh,* the main characters act out a violent crime while a television broadcast of *Archibaldo de la Cruz* plays in the background, effectively creating a story within a story. At the same moment when Almodóvar's polychromed protagonists are in the process of violently destroying each other, Buñuel's black-and-white Archibaldo drags a bosomy female mannequin from his living room to his pottery workshop so that he can cremate "her" and take pleasure in watching her fiery destruction through the window of his kiln. The disturbing combination of auto-erotica and murderous fantasies in Buñuel's 1950s film noir plays a critical role in disrupting normative definitions of generic form and content that, at first glance (especially through today's postmodern lens), Buñuel may appear to deploy without question. Moreover, by placing *Archibaldo de la Cruz* in the background of the plot of *Live Flesh,* and by consuming the same dummy for visual fuel almost forty years after the "fact," Almodóvar establishes his own movie as a filmic object, a gesture that is in keeping with the self-reflexivity of Buñuel's surrealist origins.

Nevertheless, in looking more closely at *Archibaldo de la Cruz* and *Live Flesh,* we find that both tales are highly embedded narratives framed by historical events. In Buñuel's film, Archi's life is changed forever one night when a battle of the Mexican Revolution is fought on the threshold of his boyhood home. In Almodóvar's movie, Víctor's life begins the night his mother gives birth on a city bus in the midst of a "state of exception," that is to say, an official suspension of civil rights during the death throes of the Franco regime. In each case, the protagonist's life is shaped by the flow of history, a process that is always beyond his reach and control. Furthermore, as both characters age, they demonstrate an inability to mature. Like Buñuel's middle-aged antihero, Almodóvar's young protagonist, Víctor, is sexually inexperienced in his contact with women, and the youth uses the televised film to help him establish more concrete (albeit violent) ways of fantasizing about women. As Archi and the lens of Buñuel's camera gaze at the female mannequin's legs while *Archibaldo de la Cruz* plays on TV, Víctor shifts his gaze to his love interest's legs resting on the couch opposite him. Like Archi, he has to rehearse (hence Buñuel's original title, *Ensayo de un crimen* [Rehearsal of a crime]) in an attempt to bring his fantasies to fruition. In this regard, Almodóvar deploys Buñuel's film as an intertextual referent that both represents and stimulates voyeuristic fantasies in which the subject fetishizes garments, body parts, and other physical images. Neither director attempts to vilify such fantasies in themselves, but each uses them in order to explore extreme mental states, such as psychosis and obsessional neurosis, that resulted from emotional traumas related to historical situations. Archi's story is rooted in the temporary displacement of his well-heeled family during the Mexican Revolution; Víctor's is connected to the economic, social, and political retardation his family shared with many in Spain at the end of the Franco regime.

As one might expect, Almodóvar's respect for Buñuel extends to numerous films in addition to *Archibaldo de la Cruz*; and so, the junior director employs certain actors in *Live Flesh* who hark back to the senior filmmaker's long-running desires and futile attempts to fill the void with mystical objects. For example, as José Arroyo explains, Almodóvar employs Ángela Molina, who "is famous for playing talented low-class women with high-class looks. Almodóvar uses a vibrant and youthful painting of her in full flamenco regalia to remind us of the beauty she was in . . . *That Obscure Object of Desire*."[11] Together with Molina, who Smith calls "the southern temptress" and "Buñuel's icon," there is also "Liberto Rabal . . . the real-life grandson of Francisco Rabal, the veteran actor who played the youthful *señorito* [a young member of the land-holding oligarchy] in Buñuel's *Viridiana*."[12] Furthermore, in *Nazarín* (Nazarene, 1958) the senior Rabal played a priest who takes his vows of poverty quite literally during prerevolutionary Mexico. Certainly, in these instances, the referent becomes the directors themselves, that is to say, a metastatement about Buñuel and Almodóvar's filmmaking. But on closer examination, these intertextual references—that is, the use of the same actor (Molina) or an actor who reminds us of a famous relative (Francisco Rabal)—also recall the roles the actors played as well as the specific historical contexts of their tales in which Buñuel critiques class relations in both Spain and Latin America.

Although such gestures could be ascribed to a postmodern aesthetic, their origins are deeply rooted in modernity. Indeed, the exploration of desire does not have to be void of history, and in this respect Almodóvar's self-reflexivity has an objective. According to Smith:

> Newly mature, Almodóvar can live up to the comparison [to Buñuel]; indeed, his subtle take on Spanish institutions in *Live Flesh* makes Buñuel's casual misogyny and anticlericalism seem as mechanical as the mannequin in the sequence from *Archibaldo* we glimpse on Elena's [Víctor's love interest] television screen.
>
> It is perhaps surprising, then, that this stylistic circularity gives way to a faith in linearity, even progress: both Spain and Víctor are decidedly different and better at the end of the film.[13]

Given that Almodóvar chooses to pay tribute to *Archibaldo de la Cruz*, Smith is perhaps too dismissive of Buñuel in the above comment; otherwise, he is quite accurate in his argument. It would, therefore, behoove us to delve more deeply into why Almodóvar's elaborate celluloid quotation from Buñuel's film noir coincides with the younger filmmaker's sudden shift in the way he elects to balance form and content.

To this extent, a turn toward the modern does not mean that the films of either director are easily decoded and reduced to a single meaning. Due to the tension between form and content that characterizes the medium they both share, even the most brilliant of psychoanalytically informed scholars can stumble over a carefully laid filmic object, including a bogusly benign pair of stilettos. As late as the early eighties, without access to a manuscript copy of an unpublished screenplay or without a print of a film and a private screening room with which to view the same reels over and over, it stands to reason that any public spectator could commit "errors" in trying to recall major events of a movie, not to mention the subtlest of details. For example, François Truffaut makes a significant mistake in his plot summary of *Archibaldo de la Cruz* by killing both members of a couple when, in fact, only the female lover dies.[14] Nonetheless, discrepancies still appear in recent criticism of *Archibaldo de la Cruz*, attesting to how well Buñuel's cinematography continues to challenge the perceptual apparatus of today's viewers. Evans, for instance, provides an excellent reading of the film, but while describing one scene in particular, he sees blood on the *leg* of little Archi's dead governess when in actuality it originates from a wound on her neck.[15] Evans's false memory is probably based on a repetition of the same basic scene later in the film when the cinematographer superimposes a large quantity of blood over the camera frame, effectively staining our view (and memory) of the governess's shapely hips and legs. Through our attempts to remember and retell the story, we reveal our own subjective desires and implicate ourselves as voyeurs who are, therefore, complicit in the crimes of an obsessional neurotic. More importantly, the split between modern and postmodern aesthetic visions also becomes blurry since, in this instance, the movie director produces seemingly postmodern effects through the spectator's difficulties in reading a modern story told through modern techniques of filmmaking.

The appropriation of the works of others during the creative process, often resulting in a self-referential form of pastiche, is another aspect usually attributed to postmodernism; however, this is a potentially misleading feature when we look at modernism's long tradition of parody and satire. Stemming from the self-referentiality of his surrealist origins, Buñuel quotes the works of colleagues as well as his own earlier projects in *Archibaldo de la Cruz*, especially when it comes to certain objects that embody Archi's more violent forms of desire. Without attempting an exhaustive survey of specific works and literary traditions that influenced Buñuel (for instance, picaresque narratives, wife-murder tragedies, detective fiction), there is one movie that stands out, the likes of which *Archibaldo de la Cruz* both mirrors and parodies: Alfred Hitchcock's *Spellbound* (1945).[16] Of particular note is the use of surrealist art as a means of visually representing the protagonist's pathological behavior. In this respect, Salvador Dalí, Buñuel's contemporary and former friend, made a significant contribution to Hitchcock's work with several minutes of footage for Gregory Peck's high-anxiety dream sequence.[17] Indeed, throughout *Archibaldo de la Cruz,* Ernesto Alonso (adult Archibaldo) experiences trances that echo Peck's delusional spells, effectively recreating the subgenre of Hitchcock's psychothriller.

However, *Archibaldo de la Cruz* is also intriguing for what the director refuses to quote from *Spellbound*; Buñuel completely ignores Dalí's participation in Hitchcock's film and focuses on the English director's more brilliant moments, which, when filtered through Buñuel's lens, ironically recall Buñuel and Dalí's collaboration on their first silent movie *Un Chien andalou* (1928). For example, when Hitchcock's protagonist sleepwalks one night, he menacingly carries a straight razor into the study of an old psychoanalyst who offers him a glass a milk. While Archi visits socialite Patricia Terrazas in her apartment late one night, the young vixen slips into something more comfortable, and then she goes to the kitchen for refreshments. Equipped with a straight razor, Archi takes a seat, smoke and organ music envelop him, and we are privy to one of his elaborate fantasies in which a white blur floats in the middle of a black, smoky frame. Then a zombielike Patricia dressed in black comes into focus, carrying a tray; the white image, as it turns out, is a solitary glass of milk, recalling little Archi's beverage the night the Mexican Revolution violently interrupted his life. Then, the scene advances quickly, displaying a rapid succession of close-ups and profiles over a black background, accompanied by smoke, music, and a disembodied dialogue of seduction. In the sequence the camera quickly cuts to 1) Archi's eyes; 2) Patricia, as someone approaches her in the dark; 3) two masculine hands moving up Patricia's back, a straight-edge razor in the right hand, the left used to lift her hair and reveal her neck; 4) Archi's eyes; 5) Patricia's face, a silent scream coming from her mouth while she appears to fall backwards; 6) a bloody razor in a man's right hand; and finally, 7) a cross-over which returns us to a view of Archi seated in Patricia's living room. The music also fades, and Archi wakes from his trance in a cold sweat.

Does this filmic quotation within a filmic quotation represent a "pure" metalanguage in the poststructuralist/postmodernist sense? Archi's hallucination is most certainly disturbing and frightening, reminding us of Peck's sinister handling of the straight razor as he stalks other characters in *Spellbound.* In addition, Archi's delusional moment recalls the chilling opening sequence of *Un Chien andalou* in which a man with a straight razor in his hand mechanically slices a woman's eyeball. With the utmost respect paid to Hitchcock, and without directly parodying Dalí's involvement in *Spellbound,* Buñuel skillfully reestablishes his authority over his old rival Dalí. Thus, the object could be read as a metastatement about who is the better surrealist; that is, the nightmarish scene in *Archibaldo de la Cruz* reminds spectators of the history of the artists themselves in order to determine who constitutes the real father of surrealist cinema. However, in the context of Zizek's claim that "*there is no language without object,*" Buñuel's cinematic reference to Dalí also could be read as an attempt to transcend the postmodernist framework. To this end, personal and professional animosities can be inscribed within the rubric of greater historical conflicts.

In an obituary for Buñuel written in 1983, Paul Mayersberg claims that Buñuel's art resembles that of surrealist poet Federico García Lorca more than that of Dalí.[18] Certainly, the comparison holds true in terms of the personal and political trajectories of the three. Whereas Lorca was executed by the fascists during the Spanish civil war, Dalí remained in Francoist Spain and gained international fame. For his part, Buñuel went into exile twice: first, by escaping Europe for the United States after the fall of the Spanish Republic and on the eve of the Nazi invasion of France; second, by fleeing the United States for Mexico during the red scares of the McCarthy era in the midtwentieth century. In light of life-changing events, which demonstrate why history and ideology matter, is it any wonder that much of Buñuel's oeuvre was critical of twentieth-century society and its deadly political machinations? As Almodóvar might tend to agree, although many works from Buñuel's Mexican period look like the standard feature-length, major-release movies that were commercially produced for Western audiences, they are, in fact, highly satirical of the moral and political conservatism which characterized the 1940s and 1950s. Indeed, the filming of violent sexual *fantasies* are rather tame when compared to the *realities* of our age; however, Buñuel uses such sequences to upset bourgeois sensibilities, thereby disrupting normative standards of generic form and content in films like *Archibaldo de la Cruz.*

It could be argued that Buñuel's work has numerous postmodernist characteristics because the director is never a strict adherent to any canon, surrealist or otherwise. However, the negation of all forms of narrative, including history, is not evident throughout Buñuel's long career. Even when the surrealists declared war on the anecdotal, denying the subject's capacity to say what s/he means through language, Buñuel retained narrative threads in his work. As for his later career, Guy Scarpetta explains:

> Là où le surréalisme, au-delà de la simple contestation du réalisme, en vient à suspecter toute forme de récit romanesque, Buñuel, lui, n'hésite pas à définir de façon extrêmement "réaliste" le cadre social de ses fictions (*Le journal d'une femme de chambre*, *Tristana*, *Le charme discret de la bourgeoisie*), et s'inscrit dans une tradition narrative explicite, celle du récit picaresque (*La voie lactée*, *Le fantôme de la liberté*).[19]
>
> [In contrast to surrealism, which goes beyond simple refutation of realism to profound suspicion of any form of novelistic convention, Buñuel does not hesitate to define the social context of his fictions in an extremely realistic manner (*Diary of a Chambermaid*, *Tristana*, *The Discreet Charm of the Bourgeoisie*) and to inscribe himself in an explicit narrative tradition—that of the picaresque (*The Milky Way*, *The Phantom of Liberty*).]

With the rise of fascism during the first half of the twentieth century came a concerted effort by totalitarian states to rewrite history and to eliminate people who served as reminders of certain historical truths. It would appear, then, that Buñuel's initial aversion to his colleagues' abandonment of storytelling is linked to an ideological investment in history rather than a simple unwillingness or an intellectual incapacity to fulfill the tenets of a surrealist manifesto. In part, Buñuel's deployment of the picaresque relies on its long-standing tradition, if not generic obligation, to provide a critique of society and its history. To this extent, there is a striking contrast between Buñuel's modernist surrealism and Dalí's aesthetically nonreferential behavior, particularly in light of the latter's indifference to fascism.

If, however, it were possible to discuss "art for art's sake," if we could remove the object of art from its historical and political milieu, then psychoanalytically and aesthetically speaking, Buñuel and Dalí are very similar in their treatment of the object of desire. Zizek cites films by Buñuel, including *Archibaldo de la Cruz* (*SO*, 194), to explain the "sacred/forbidden, empty place in the Other" (*SO*, 195), a concept Dalí also portrays well. However, Zizek's comments on poststructuralism are also enlightening for our discussion on the attempt to divide the modern from the postmodern:

> the post-structuralist procedure *par excellence* is not only to search in purely literary texts for propositions containing a theory about their own functioning but also to read theoretical texts themselves as "litera-

ture"—and more precisely, to put in parentheses their claim to truth in order to expose the textual mechanisms producing the "truth effect." . . . Lacan always insists on psychoanalysis as a truth-experience: his thesis that truth is structured like a fiction has nothing at all to do with a post-structuralist reduction of the truth-dimension to a textual "truth-effect." (*SO*, 153–54)

In this context, Buñuel and Lorca seem to share a political investment in storytelling, but Dalí's static visual medium appears to break down the anecdotal—to destroy history itself—while refusing to acknowledge, as Zizek also states, that "the subject must undergo the experience of how the ever-lacking object-cause of desire is in itself nothing but an objectification" (*SO*, 195). No matter how abstract, Buñuel's moving images and Lorca's dramas thrive on the linear and the anecdotal. Both are grounded in (hi)story; to this extent, the straight-razor sequence in *Archibaldo de la Cruz* demonstrates how Buñuel embraces the surrealist aesthetic of his youth but rejects some of surrealism's philosophical underpinnings. Is it, in fact, possible for us to arrive at the truth? In his own quirky way, Buñuel seems to say "yes."

In deploying a similar cinematic language of desire to the one spoken by Buñuel, Almodóvar is equally capable of situating people's personal stories within an historical context. Indeed, Almodóvar's quotation from *Archibaldo de la Cruz* achieves a result that parallels Buñuel's fantastical exploration of desire. As the judge states when he exonerates Archi of all crimes, "El pensamiento no delinque" [Thought is not a crime]. In other words, the human imagination is as close as we will ever come to true freedom. Despite the fact that desire, according to Lacan, escapes symbolization, its pursuit provides us with our only glimpse of what Buñuel calls "the phantom of liberty," a phrase he selects for the title of a film in 1974. Through his written comments about Buñuel's Mexican productions, Almodóvar explains how we are all prisoners of desire; however, he also implies that fantasies, no matter how frustrating, are of our own making:

The first scene of *Él* . . . takes place in a church; a priest washes the feet of the faithful in an Easter ceremony. From the protagonists [*sic*] point of view, the camera shows the feet of some of the penitents until it reaches two black high heels from which a pair of beautiful legs emerge. The owner of said extremities will become the owner of his heart, and the prisoner of the most paranoid and cruel jealousy.[20]

And so, in *Live Flesh*, Almodóvar literally focuses on the same objects Buñuel studies, such as the mannequin's wax leg in *Archibaldo de la*

Cruz, which falls off while Archi drags her to her "death." But by picking up the leg where Buñuel's protagonist dropped it, is Almodóvar proclaiming himself a modernist?

In part, a definitive answer would require a peek into the future. Certainly, his current trajectory—including the release of *Todo sobre mi madre* (All about my mother, 1999) in which Almodóvar pays homage to the modernist romantic vision of Lorca, Bette Davis, and Tennessee Williams—indicates that he has struck an intriguing balance between modern storytelling and postmodern imagery, an equilibrium as original as that of his early years. We also will need to examine the entirety of his career in retrospect, comparing his craft to that of other filmmakers, particularly Buñuel; in other words, it will be necessary to place Almodóvar within the context of his cultural, artistic, and intellectual heritage. In the end, Almodóvar—like Buñuel, whose little Archi debuts in *The Criminal Life of Archibaldo de la Cruz* wearing his mother's dress, shoes, hat, and corset over his pajamas—never loses faith in the phantom of liberty and the pursuit of those obscure objects of desire that help us to fill the void and to express what words alone cannot fully convey. However, unlike Buñuel, who never seems to forget that his directorial career was almost ended by historical circumstances, Almodóvar has to remind himself in *Live Flesh* of how fortunate he was to begin his career in post-Francoist Spain and to continue his filmmaking during a time of relative peace, relative prosperity, and relative artistic freedom. Therefore, it still remains to be seen whether Almodóvar will continue to favor content over form by acknowledging the materiality of objects as referents to history and real events. If Buñuel's relationship to surrealism is any indication, it would appear that Almodóvar is following a parallel course by embracing postmodernism and some aspects of poststructuralist thought without abandoning storytelling and its ability to help him communicate historical truths.

Notes

1. Pedro Almodóvar, "About the Production, in His Own Words," *Live Flesh*, 9 January 1998, 3 March 1998 <http://www.mgm.com/live_flesh/story_pedronote2.html>.

2. For a detailed study of objects appearing in films by Buñuel, particularly in the context of their use as poetic devices, see Antonio Monegal Brancós, *Luis Buñuel de la literatura al cine: Una poética del objeto* (Barcelona: Anthropos, 1993).

3. Among countless studies employing psychoanalysis about Spanish-language cinema, the following monograph on Buñuel stands out: Peter William Evans, *The Films of Luis Buñuel: Subjectivity and Desire* (New York: Oxford University Press, 1995).

4. Beginning in Madrid at the Museo Nacional Centro de Arte Reina Sofía (16 July to 14 October 1996) and moving to Mexico City at the Museo del Palacio de Bellas Artes (4 Decem-

ber 1996 to 2 March 1997), the art exhibit entitled "¿Buñuel!: La mirada del siglo" (Buñuel!: The gaze of the century) was a tour de force in terms of its focus on the repetition of images in Buñuel and his contemporaries. However, for those unfamiliar with his work or the decades in which he worked, the exhibit offered little in the way of explaining Buñuel's art and its relationship to twentieth-century history and politics.

5. For an excellent discussion of Almodóvar as postmodernist, see Paul Julian Smith, *Laws of Desire: Questions of Homosexuality in Spanish Writing and Film, 1960–1990* (London: Oxford University Press, 1992).

6. Slavoj Zizek, *The Sublime Object of Ideology* (London: Verso Books, 1992). Hereafter *SO*, cited in text.

7. Evans, *Films of Luis Buñuel*, 2.

8. Paul Julian Smith, "Absolute Precision," review of Almodóvar's *Live Flesh*, *Sight and Sound* 8, no. 4 (1998): 9.

9. Ibid., 7.

10. This, of course, is not the first time Almodóvar has paid tribute to Buñuel in a film, but scenes in *Live Flesh* may be his most direct references. For example, the prolonged confinement of the protagonists in a small space throughout much of the film *¡Átame!* (Tie me up, tie me down, 1990) is highly reminiscent of one of Buñuel's most famous Mexican productions. As Smith states, "Almodóvar genuflects . . . to Buñuel's masterpiece of inertia and confinement *El ángel exterminador* ('The Exterminating Angel,' 1962)" (*Laws of Desire*, 207).

11. José Arroyo, "*Carne trémula*," review of Almodóvar's *Live Flesh*, *Sight and Sound* 8, no. 5 (1998): 51.

12. Smith, "Absolute Precision," 8.

13. Ibid., 8.

14. François Truffaut, *The Films of My Life* (New York: Simon & Schuster, 1978), 265.

15. Evans, *Films of Luis Buñuel*, 102, 104.

16. For discussions of Buñuel's literary heritage, see my essays "Quixotic Desire and the Avoidance of Closure in Luis Buñuel's *The Criminal Life of Archibaldo de la Cruz*," *Modern Language Notes* 114 (1999):269–96 and "Through the Looking Glass: Reflections on the Baroque in Luis Buñuel's *The Criminal Life of Archibaldo de la Cruz*," in *Echoes and Inscriptions*, ed. Barbara Simerka and Christopher Weimer (Lewisburg, Penna.: Bucknell University Press, 2000).

17. While *Spellbound* is a truly unique film for its time, as a psychological thriller it seems trite today because it strings together symbols of Freudian analysis in order to create a dramatically serious narrative about why Peck's character is suffering from amnesia. In this respect, *Archibaldo de la Cruz* is at times a comic parody of the relationship between Hollywood celluloid and psychoanalysis, a dynamic Almodóvar also uproariously satirizes during the climax of his motion picture *Laberinto de pasiones* (Labyrinth of passion, 1982).

18. Paul Mayersberg, "The Happy Ending of Luis Buñuel," *Sight and Sound*, no. 4 (1983), 258.

19. Guy Scarpetta, "Buñuel: surréaliste ou baroque?" *Art Press*, no. 129 (1988), 64. My translation.

20. Almodóvar, "About the Production."

1918: García Lorca at the Crossroads

Luis Fernández Cifuentes
Harvard University

THE same terms that Luis García Montero devoted to Lorca's earliest—and only recently published—poetry, may apply to Lorca's first published book, *Impresiones y paisajes* (1918). A critic and a poet, García Montero writes at (and, in part, about) the belated point in time when the old juvenilia manuscripts abandoned the private sphere of family archives to join the public sphere of print culture—*print cult* indeed, in the case of Lorca's fervent readers and scholars. In those adolescent, often embarrassingly clumsy lines, which the mature Lorca never thought of publishing, he gave voice to intimate confessions but only by the voices that were not his own, in a paradoxical—and, again, awkward—combination of ventriloquism and self-disclosure. García Montero writes, "García Lorca's poetic world is an absence, a palpable void over these drafts. What there is of him is precisely that which he shares with others"; "García Lorca's youthful poems are written through the voices and shadows of Rubén Darío, Juan Ramón Jiménez, Miguel de Unamuno and Antonio Machado."[1]

In the case of *Impresiones y paisajes,* a book of prose published by the young author with his wealthy father's help, the voices of Azorín, Valle-Inclán and, perhaps to a lesser extent, Baroja and Gutiérrez Solana, should be added to the list. Thus, the most prominent figures of the older generation can be overheard in a brief book that is densely populated with multiple voices, apparently suffocating the author's own literary identity. At first sight, an informed reader may be tempted to dismiss the work as a mere anthology of a schoolboy's exercises in imitation, a naive tribute, at best, to the ruling generation. A scholar bent on erudition may be compelled to expose the connection between so many of Lorca's expressions and the models of his predecessors. Such a scholar may point to the emphatic resonance of Unamuno's "these excursions are not only a consolation, a rest, and a learning experience; they are also and perhaps above all, one of the best ways of gaining attachment and love for the homeland,"[2] in Lorca's "in these sentimental walks full of unction for the Spain of the warriors, the soul and the senses enjoy everything and become intoxicated with new emotions that can only be learned here."[3] The scholar may detect the *common place* that Galicia turns out to be in both Azorín's "all is green under the fine rain from a cloudy sky"[4] and Lorca's "the rain falls silently and slowly over the sweet green of the land" (*I*, 247). It is also difficult to miss the explicit, hardly equivocal relationship between *Jardines lejanos*, Juan Ramón Jiménez's collection of poems published in 1904 (and perhaps also *Las hojas verdes*, from 1908), and Lorca's "nebulous gardens that make Juan Ramón Jiménez, the great poet of the mist, suffer" (*I*, 187). Even the book's very genre refers the reader to the older intellectuals; *Impresiones y paisajes* is a travelogue of sorts,[5] nearly unique in the prose production of Lorca's generation, but much like those travelogues so frequently and significantly present in the writings, as well as in the readings, of the Generation of 98.

And yet, too many traits in Lorca's opinionated first book do not entirely conform to that image of submission and deference. In other words, certain discrepancies, contradictions, juxtapositions, and reversals create frequent disturbances in Lorca's prose that speak less of acceptance and imitation than of resistance and displacement; less of suture and continuity than of rupture and differentiation. In this light, the apparently premature publication of *Impresiones y paisajes* may not represent just the whim of a spoiled adolescent (Lorca was eighteen and nineteen when he wrote most of its pages), but rather a gesture of defiance toward the elders of the literary and intellectual community. I wish to argue that *Impresiones y paisajes* should be read as the rare documentation of a break between generations, as one of those unusual literary sites where the informed reader can observe historical transitions taking place, trends and tendencies coming together, coming apart, or both. I trace this performance of dissidence in three different qualifying factors of the book: choice of stylistic resources, incidences of the traveler's social class, and redistribution of the national map.

Ian Gibson, with the skill and thoroughness of a professional sleuth, has uncovered the presence of a radical breach within the realm of adopted voices that appear to crowd Lorca's text. He has also established the connection of such a breach with one of the harshest, most significant transitions in the poet's life. The trips described in *Impresiones y paisajes* were organized by Martín Domínguez

Berrueta, Lorca's professor of Theory of Literature and Art. Berrueta (1869–1920)—five years younger than Unamuno, six years older than Machado—was himself a full-fledged, paradigmatic member of the Generation of 98. Gibson writes that time and again, as was to be expected, "we hear the voice of Martín Domínguez Berrueta" in the pages of *Impresiones y paisajes,* which Lorca originally planned to publish with a prologue by the old professor. However, on the one hand, the prologue was never requested, and, on the other, a few of the most compliant paragraphs of the text—on Pereira's sculpture of San Bruno, at the Cartuja de Miraflores, which had also been the subject of one of Berrueta's characteristic articles—were entirely rewritten after Lorca's groupies heard him read them and concluded that "it was Berrueta's voice which spoke through them, not that of Federico." Lorca's travel companions considered the new, contentious paragraphs a "brutal betrayal" to Berrueta. They appear to have been also the reason for "an abrupt and definite break between professor and disciple."[6]

It was not only a matter of dissidence by reversal, though. This event is precisely an avatar of a more pervasive, if less painful, phenomenon that characterizes the entire set of literary voices in Lorca's book, even when limited to the chapters on Burgos and its Cartuja. We may call it a matter of conflicting affiliations. However inconsistent or disperse the book may sound in its intent to reconcile such different voices as those of Berrueta and Juan Ramón, Unamuno and Valle-Inclán, it would still be a rather homogeneous product if it simply attended to a single, more or less unified, national generation. The fact is that other, more alien, voices contribute to make the book a more heterogeneous, even hybrid, aggregate. This condition can be summarized as follows: two sides of a long-lasting, never-resolved controversy have tried to establish either the association or the dissociation of two hegemonic literary trends of turn-of-the-century Spain, "modernismo" and "98." The first movement can be identified, for good or, more frequently, for worse, with the "frivolous," "sensual" aestheticism of a number of cosmopolitan Latin American writers, headed by Rubén Darío. The movement called "98" can mostly be identified with the national writers who, devastated by the so-called disaster of the Spanish American war, focused their somber attention on the old towns and mostly bare landscapes of Castile. Naturally, both groups favored travelogues—more often called "crónicas" [chronicles] for the *modernistas,* "ensayos" [essays] for the *noventayochistas* [the ninety-eightists]—which reflect their differences perhaps more acutely than any other genre they cultivated.[7] *Impresiones y paisajes,* an heir of that genre, constitutes one of those scarce scenarios where both trends appear side by side in equivocal coexistence: it seems to serve those who deem the trends reconcilable as well as those who only note the terms of their opposition. More likely, it serves neither movement, but rather their conflicts and frictions. Burgos is a case in point, a site particularly favored by Berrueta, and the longest stop of the entire set of trips. In letters to his parents, Lorca typically declared: "Don Martín is the master here. As soon as we arrived and got settled, Don Martín told me the study plan I should follow and I have started my homework. This is marvelous and worthy of study because of the quantity of past history that is contained in all the stones."[8] In the corresponding fragment of *Impresiones y paisajes*, however, the commanding, enthusiastic voice of Berrueta or, for that matter, Unamuno,[9] is heard less so than the harsh, disappointed, critical tones of an Argentinean chronicler such as Manuel Ugarte. A few years earlier Ugarte had described Burgos's cathedral as "a monster nestled in the shadows, behind which the moon hides and reappears." Lorca's description is uncannily similar, with just a hyperbolic touch that is closer to Darío de Regoyos:[10] "the gothic monster of its cathedral, the work of a giant silversmith, cut over a triumph of a purple color" (*I*, 38). They similarly coincide in their evaluation of the religious atmosphere in the town and its churches. Here is Ugarte's:

> From where come these chimerical beings, which slide with small steps, making genuflections before the altars, and disappear in the sea of hunched bodies, which block the center of the nave? What mentality do they have? Their features do not have the repose and placidity of the just, but instead tortured and anguished expressions, as if they were terrorized by the obsession of hell . . . Modern man drowns in this vanquished world, where everything seems to be annihiliation, sadness, and eternal death.[11]

Lorca agrees, although he is not quite as explicit about his condition of "modern man" vis-à-vis the ancient town:

> This ascetic and silent countryside has the charm of a painful religiosity. The eternal hand poured only melancholy over it. All things express a formidable bitterness and desolation in their forms. The vision of God in this landscape is one of immense fear. Everything is overwhelmed, frightened, flat. The poor soul of the people manifests its anguish in their speaking, in their slow and grave gait, in their fear of the Devil, in their superstition. (*I*, 39)

This form of dissidence by association with another, foreign, cosmopolitan legacy finds at least two other forms of textual articulation in *Impresiones y paisajes.* First, Lorca appears to have discovered a special affinity with one particular feature that the *modernistas* adopted from French Parnassianism: the general supremacy of music and musicality or, in Verlaine's revered motto, "de la musique avant toute chose" [music above all]. Among the poets of the Generation of 98, Unamuno questioned such supremacy with significant acerbity. For example, in 1916, in regard to Darío's poetry Unamuno wrote: "The eternal and intimate concerns of the spirit overwhelmed him and they inspired his most profound, most intimate, and greatest poems. Not those guitar pieces that are usually cited when his poetry is mentioned, about 'the princess is sad,' 'what is wrong with the princess,' or about the 'the wing elevates from the light fan,' that don't go beyond slight tickling of a frivolous acoustic sensuality."[12] In *Impresiones y paisajes,* music may be the most emphatic and distinctive trademark of the author who was known by his travel companions as "the musician."[13] Lorca opens the book with a fervent dedication to his music teacher. The resort to musical terms and correspondences in his descriptions occurs on virtually every page, often more than once.[14] It is, however, in the direct reference to musicians and their compositions that he appears to strike not only a more personal note but also a more discordant one. In the initial "Meditación" about the old Castilian towns that formed the main stops along the itinerary, he concludes: "Oh cities of Castile, upon contemplating you so severe, my lips say something of Handel" (*I,* 21). This appears to be not only a pedantic statement, but also a rather opaque one. The uninformed reader may take it as just one more musical reference, while the more informed reader wonders what could possibly be the connection between the baroque, often grandiloquent, distinctly "international"[15] composer and the austere medieval towns, emblems of the new *castizo* [pure] nationalism. Was García Lorca simply trying to claim a distance between those places and his own sensibility, or, better yet, between the intellectuals of 1898, who had identified with those towns, and his own approach to the geographical and cultural landscape? Historical and literary erudition characterizes and even determines not only the travels but also the travel writings of the Generation of 98.[16] In *Impresiones y paisajes,* Lorca—however well instructed by Berrueta's course—mentions far fewer writers and painters than one will find in Unamuno or Azorín's *models,* but he adds a number of composers that had no place or function in their travelogues. Was he trying to counterbalance his predecessor's imposing library with his unusual (if not extraordinary) erudition of records and scores?[17] Lorca's description of San Pedro de Cardeña appears to support both hypotheses: the chapter ends with a sharply contrasting juxtaposition of certain external "iron voices over the fields" (an obvious reference to *Poema de Mío Cid,* so much an ideological and literary staple for the Generation of 98) and "the sobbing of an afternoon song by Schuman that painfully passes through my soul" (*I,* 61). The reader could draw similar conclusions from a more arrogant anecdote recorded in great detail: when Lorca learns that the old organist of Santo Domingo de Silos knows nothing but Gregorian chant and its organ accompaniment, he shows off—rather condescendingly—his knowledge of the other musical instruments. Then, both as a lesson to the monk and as indication of his own "modern" musical sensibility, Lorca mentions Beethoven (who did not compose anything memorable for organ and was, in 1917, the object of serious critiques by the likes of Stravinsky) and finally, he demonstrates the caliber of Beethoven's music by playing (at the organ!) the first bars of the "allegretto" from the seventh symphony (*I,* 99).

The relationship Unamuno traces immediately between musicality and sensuality ("frivolous acoustic sensuality") in the passage quoted above brings to the foreground a second articulation of this same form of musical dissidence. Unamuno had written much earlier, in the essays that would later be collected under the title *En torno al casticismo* (a major influence on *Impresiones y paisajes*): "Castilian realism is more sensitive than sensual, without imaginative refinements, and with a chaste background. It smacks more of tavern than of bordello, and when is taken to extreme, it tends, even in obscenity, more towards the gross than to the libidinous."[18] Lorca would generally associate himself with sensuality and its "imaginative refinements" in more than one way. For one, the formal sensuality of his descriptive register is primarily attuned to the pleasures of melodious sounds, smooth textures, and luxuriantly colorful spectacles. Second, the words "sensual" and "sensualidad" [sensuality]—as well as an entourage of related terms, like "voluptuosamente" [voluptuously], "desfallecimiento" [fainting], "lánguido" [languid], etc.—appear in this text with significant frequency, and lacking derogatory connotations. Moreover, this sensuality is not geographically or socially circumscribed: it registers Granada's "enormous sensuality" in the moonlight (*I,* 169) as well as the "sensual souls" that are at home in Lorca's descriptions of varied, adored gardens (*I,* 181). Sensuality blooms among the Carthusians of Miraflores (*I,* 52) and the nuns of an unidentified convent (*I,* 232), as well as it defines the lips of "coarse and big beautiful young girls" who take a stroll Sunday evening in a generic Spanish village (*I,* 238). Lorca could certainly have located the obscenity and crude lust Unamuno accredited in typical Spanish sexual behavior, but he would not be inclined to generalize as much as he was apt to display its more repugnant, horrifying extremes (see, for example, his "Canéfora de pesadilla").[19] As an alternative, Lorca does not necessarily propound chastity or inhibition, but rather he glorifies the profane beauty of an impassioned sexual love that moves away from the habits of the ignoble brothel toward Bradomín's elegant, histrionic, sentimental perversions: "A garden is a tabernacle of passions, and a grandiose cathedral for beautiful sins" (*I*, 180). Finally, Lorca's—and Lorca's generation's—different take on sensuality and sexuality are significantly displayed here in his approach to the body and the body's erogenous zones. He seems to have overcome the (more or less religious) fear of the body or, at the very least, the discomfort that the naked body—feminine but, especially, masculine—generated among the sober and somber members of the Generation of 98: "The women of the small towns bathe in the river. They shriek with pleasure at feeling the coolness of the water's licking their stomachs and breasts. Men, like fauns, hide in the bushes to see them naked. A group of naked boys with bronzed skins bathe in the creek, and upon emerging, they roll around in the road's hot dust with pleasure" (*I*, 226). Gibson has noted Lorca's "obsessive" battle against sexual repression, which he carries on with particular vehemence in his repeated critique of monastic life throughout the book.[20] Today's reader may find the critique arrogant and self-righteous, at least more so than the deceivingly objective quality of direct descriptions of the body, naked and otherwise, clearly detached from overwhelming religious environments: "The dark, ruddy, and bold young girls would proudly show off their brightly colored silk blouses, their fake gold chains, their enormous and tremulous breasts" (*I*, 238). A reader may even conclude that it is the repeated juxtaposition of the sacred and the profane, the subjective critique and the objective description, that brings out once more Lorca's difference. The traveler himself explicitly broached that kind of combination: "One must be religious and profane. One must reunite the mysticism of a grave gothic cathedral with the wonder of pagan Greece. One must see it all, feel it all" (*I*, 12).

The use and abuse of musical knowledge in the text is instrumental not only in displaying the æsthetic positioning of the young traveler vis-à-vis the older generation, but also in exposing a social distance, indeed a class fissure between generations. The episode involving the "poor" old organist and the young musician (*I*, 98), is hardly the only example; the priggish complaints about a rustic coachman who will not sing "some noble Castilian song" instead of "a frightful *cuplé* [variety song] tinted with ugly Madrilenian arrogance" (*I*, 67), and the long-winded, implacable scorn of the "poor" village musicians and "the failures of air and technique" of their Sunday concert (*I*, 239), also serve to illustrate this distance. What may appear to simply expose legitimate aesthetic disappointments, articulates, particularly in the last incident, an unbending class loftiness: "The gallant and distinguished instrument, aristocratic and literary pipe, brother of the lyre and panpipe, whose prestige was confirmed by the century of lace and clavichord . . . was held by large stoned hands covered by wrinkles and hairs that clumsily wounded the registers" (*I*, 240).[21] Strident, haughty remarks like these, brimming with upper-class disdainfulness, weigh heavily throughout the text well beyond the musical arena.

In the immediate periphery of the text, extraliterary information reminds the reader that Lorca distinguished himself during these trips not only as "the musician" and, occasionally, as a writer, but also as the unequivocally wealthiest person in the group, in stark contrast with Berrueta's Spartan travel budget.[22] His letters home are routinely punctuated with requests for additional funds, to which the father generously accedes, as he later acceded to financing the resulting book's publication. Indeed, the economics of the trip and of the ensuing publication seem to infiltrate the textual matter of *Impresiones y paisajes* in at least two different ways, both indicative of acute generational differences. On the one hand, the pronouncements of the Generation of 98 with regard not only to the Castilian landscape but its general population, oscillated between a certain idealism (as when Unamuno writes in 1889 that "at times, a villager seems like a dethroned king")[23] and cautious pessimism, be it Machado's "bad men abound in the country and the village"[24] or Baroja's "the town's young gentlemen who sang Madrilenian zarzuelas, with empty eyes, without expression or thought."[25] In either case the travelers avoided particularism, individualism, and circumstantial immediacy and reached for the national historical essence in widely representative characters. In his first book, young Lorca appears to conscientiously follow the second path.[26] Oftentimes, though, the immediate, unequivocal signs of a hierarchical class difference between the educated traveler and his object of contemplation are more prominent than the pessimistic, broad critical overviews in the text. Thus, in "Mesón de Castilla" (Inn of Castile), the traveler's notes betray not so much his original models—perhaps Azorín's "Inns," from his book *Castilla* (1912), or other masters Azorín mentions—as the perfectly polished, self-conscious child from the upper middle class of urban Granada:[27] Peasants "picked up the poor foods with their large hands of stone, and brought them to their mouths with a religious unction, and would then wipe their hands on their pants" (*I*, 33). This is simply one in a series of observations that includes the other passengers on an old stagecoach,[28] the ugly friar reading poorly out loud in the monastery's dining room (*I*, 102–3) and the Covarrubias populace who directed their prayers to two "poorly made dolls," ignoring the "superb Romanesque, Byzantine and Gothic sculptures" exhibited next to them in the village church (*I*, 73).

On the other hand, this same generational difference is evidenced in the fact that the members of the Generation of 98 typically earned most of their living as journalists throughout much of their lives: many of their publications, and in particular those collected in their travelogues, were somewhat constricted, if not altogether determined, by their condition as newspaper articles. Furthermore, they were journalists unabashedly explicit about their professional involvement and its implications. Azorín, whose *La ruta de Don Quijote* was commissioned by ABC in 1905, makes a reference to himself and his colleagues in the aforementioned work as "modest journalists."[29] Unamuno and Berrueta were, of course, professors. However, Unamuno published most of his essays in the newspapers of Latin America and Spain not only because of an economic need, but also for need of a large, hopefully malleable audience that he might not reach as immediately through a book. Gibson has written about Berrueta: His "principal talent perhaps was journalism."[30] Journalism at the time was still more concerned with opinion than with information: creating opinion among as large a constituency as possible while, at the same time, keeping it adequately informed. This is clearly reflected in at least two features of the Generation of 98's most characteristic essays: on the one hand, the number and timeliness of their bibliographical and historical references; on the other, the continuous awareness of a more or less loyal audience hungry for authorized opinions of general political and ideological relevance at each precise point in (historical) time. Thus, Unamuno would intersperse even the most lyrical of his travel notes with comments like the following (with its barely covert reference to the political turmoil of 1923): "But you have me here again, reader, in the midst of the storm, and let's see if it will carry away with it the Spain that passes: you have me here, after having stored up sun, air, loneliness, serenity—yes, serenity as well—and courage."[31]

Such a dialogic quality is not exactly absent from Lorca's travelogue, but it is often counterbalanced, if not generally canceled, by a different disposition toward writing, audience, and publication. Following Berrueta's guidance and practice, Lorca started publishing in local newspapers some of the fragments he would later include in *Impresiones y paisajes.* It was a purely mimetic gesture, though, not a form of labor or on-the-job training: Lorca, like most of the writers of his generation, would never become a full-fledged journalist, in part, no doubt, because he lacked the inclination or possibly the required skills (journalism was becoming more and more professionalized) and also in (no small) part because of his class upbringing. This difference with the older generation is already considerably salient in *Impresiones y paisajes.* An opinionated book, indeed, some of its opinions certainly appear to reflect those of his masters as well as to aim at a similar journalistic audience: "Es verdaderamente angustioso lo que pasa en España con estas reliquias arquitectónicas . . . Todo trastornado . . . pero con qué visión artística tan deplorable" [What is happening in Spain with these architectural relics is truly distressing . . . Everything turned upside down . . . but with such a deplorable artistic vision] (*I,* 131). But this kind of opinion is often, and forcefully, displaced by either a markedly æsthetic, personal whim,[32] the kind of acute elitism mentioned above, or at times, by both in a promiscuous rambling:

> The priestly duties should be carried out by women, whose hands, which are pink lilies, become lost among the white of the embroidery, hands worthy of elevating and blessing the host, lilies of truly priestly charm, and whose mouths could pose themselves over the chalice like smooth garnets of passionate purity, unique lips initiated by their beauty or by their symbolic meaning, to receive the mystic and ineffable harmonies of the celestial lamb's blood. (*I,* 94–95)

At the same time, information—bibliographical or historical, about the past or the present—appears carefully diffused in this text, certainly more so in comparison with his private correspondence at the time.[33] In *Impresiones y paisajes* there is an obvious, seemingly willful disregard for the immediacy and precision of newspaper writing, even as it was practiced in the less journalistically conscious pieces of the Generation of 98. As it turns out, Lorca made a remark years later about *Poeta en Nueva York* that can be applied generally without alteration to *Impresiones y paisajes*: "neither am I going to narrate a journey, but my lyric reaction with all sincerity and simplicity."[34]

If Lorca found himself at a generational crossroads in 1918, it is nowhere more noticeable than in the cartography of *Impresiones y paisajes*. The travel itinerary was, of course, designed for the young travelers by the old master Berrueta; which is to say, it was a product of the geographical ideology of the Generation of 98 or, in Machado's words, "Spain's towns and urban areas, where Berrueta, along with his students, seeks the live emotion of the nation's art."[35] Lorca does comply with genuine enthusiasm, as, for example, when he writes: "All of Spain's past and almost all of the present can be breathed in the solemn and august cities of Castile" (*I*, 19). And yet, the resulting book is not exactly acquiescent or submissive: the traveler, as far as we know, never strayed away from the designed path, but his travelogue did, in subtle and not so subtle ways. For one, the reader of the current, nonannotated, editions will have to resort to extraliterary sources in order to know the actual stops of the trip as well as their order and dates. The book follows no itinerary as such, the only exception being the short trip from Burgos to the Monasterio de Silos. Towns and monuments have been scattered, chronology (except for passing indications of the season) has carefully been erased as have quite a few toponyms, leaving places nameless and lost in an already vague map: where are "Un pueblo"[A town], "Una ciudad que pasa" [A passing city], "un pueblo grande" [a large town], "ciudad antigua" [ancient city], "Iglesia abandonada" [abandoned church]? They seem to be not so much actual locales as general topics for a literary exercise.

As designed by the Generation of 98, the itinerary follows mostly historical traces: space was generally considered at the service of time, whether it was Unamuno's "eternal tradition" or Azorín's "this overwhelming and terrifiyng concern with the endless succession of things through time."[36] *Impresiones y paisajes* reflects this priority at several moments: the book starts by depicting itself, in the very first sentence, as endowed with "a certain vagueness and melancholy" (*I*, 11). Immediately, the traveler exercises the "vagueness" I have explained above, and detects in almost every place a "melancholy" which is later justified as a product of time: "Time, just as it is the great destroyer and the great teacher, is the great artist of melancholy" (*I*, 112). This priority of time over space was vital for the Generation of 98's building of a Spanish national unity around the supremacy of Castile. In *En torno al casticismo*, Unamuno puts it this way:

> But if Castile has made the Spanish nation, the latter has become even more Spanish, fusing more every day the wealth of the variety of its inner contents, absorbing the Castilian spirit in another superior to it, more complex: the Spanish spirit. The regional desires, more alive by the day,

> have no other deep meaning, desires precisely felt by Castile itself; they are symptoms of the process of Spain becoming Spanish, they are prodromes of the deep work of unification.

Unamuno then goes on to trace the history that warrants Castile's authority as "the true forger of the Spanish monarchy and unity."[37] A subtle gesture of subversion against such a priority often appears to govern Lorca's book. The Generation of 27 in general, and Lorca in particular, started moving toward a significantly modern balance of priorities that would eventually result in what a postmodern geographer calls "a recomposition which enables us to see more clearly the long-hidden instrumentality of human geographies, in particular the encompassing and engaging spatializations of social life that have been associated with the historical development of capitalism."[38] I will briefly track the early manifestation of this "recomposition" in two distinctive geographical features of Lorca's book: an overwhelming presence of two opposed poles—Burgos and Granada—and a conspicuous absence of Catalonia, and in particular, Barcelona.

Time, history, an awareness of past, present, and even future,[39] are far from absent in *Impresiones y paisajes.* However, the book does not appear as intent on the recognition of historical identity as in the display of a particular spatial structure. The bulk of the book is devoted to the exploration of two very specific sites, Burgos and Granada—the former, of course, Berrueta's ancestral territory, the latter, that of Lorca. The young traveler underlines not so much the cities' roles in the historical configuration of the old nation, as their present, modern opposition on a North/South axis clearly distributive of social, cultural, and geographical signs of identity. Thus, the book avoids reproducing a physical and political geography of the nation comprised by a *center,* "Castile," and a *periphery* whose identity has been historically determined by such a center. The North/South axis does not encompass diversity within unity, as Unamuno would have it, but rather a structural difference between places of belonging and places of alienation. The travelers of the previous generation searched around Castile for original roots, locations of historical continuity, spaces of permanent belonging. Unamuno finds in Gredos "Long-lasting Spain" and cries in Brianzuelo de la Sierra: "I would like to stay here."[40] Lorca's masters were not so much traveling as returning or planning to return: the strength of their map is in its repetition. Each (national) trip must be a form of recognizing sameness and discarding or minimizing difference. Azorín delineates the perfect identity of a street in Santander, another one in

Granada, another one in Murcia: "And in many other cities you have seen a street like this street. In our old Spanish cities there is nothing more typical, more original, more cosubstantial with the race and the environment."[41] Lorca displaces that projection of unity in time and space with a more precise geographical differentiation. In and around Burgos—Miraflores, Las Huelgas, Silos, Cardena, churches, convents, monasteries, tombs—the traveler identifies, at first sight, mostly unwelcoming interiors, domains of alienation, sites to be abandoned or funereal monuments of abandonment:

> These admirable men of decision flee from the noise believing that sins are hidden in it, and have fallen into another place conducive to thought, and therefore to sin. . . . All the words of sensitive and passionate reply were once heard around these currently dead environs. . . . It is a cold, enormous, misshapen, and unpleasant church. . . . The church is deserted and damp. (*I*, 51, 59, 93)

There is no doubt, at any point, that the traveler is only that: someone who is in transit and does not belong there except for brief visits, "so when they are finished, they can leave the marvelous gamut of memories" (*I*, 21). Lorca never seems to have developed the foreseeable effect of "patriotism" that Unamuno and Machado anticipated. In Granada, on the other hand, the traveler conveys a sense of permanence and constant affinity, a general feeling of intimate property and lyrical propriety, the benefit of spontaneous knowledge and experience instead of the programmed succession of scholarly encounters. In Granada he traces the changes from hour to hour, from day to day, from season to season (*I*, 166, 154, 171–75). He shows, especially in "Sonidos," a musical affinity that was far removed from his lack of disposition to tune in to the music of the North.[42] He multiplies the gardens that—unlike the northern monastic interiors—welcome a variety of emotions and desires: "A garden is something superior, an accumulation of souls, silences, and colors, that wait for the mystical hearts to make them cry" (*I*, 180).

When Unamuno speaks of "pruritos de regionalismo" [the urge to emphasize the regional identity], I suppose he has in mind, primarily, Catalonia, which then threatened to undermine his unifying notion of a "Spanish nation." If Barcelona is so conspicuously absent from Berrueta's itinerary, it is probably due, first and foremost, to the political turmoil of the city at a time of exasperated catalanism: on 13 August 1917, after an agitated month of July, the central government declared a state of war in Catalonia, and Catalan workers called a general strike which lasted until 20 August. Lorca was then in Burgos examining religious life in Castilian monasteries. At the same time, the choice—if there ever was a choice—of Santiago de Compostela over Barcelona represented again a victory of time and history over space and modernity.[43] Berrueta's travelers were clearly going against the grain. Unamuno claimed in 1906 that "tourists and travelers only visit but a few cities, particularly Granada, Toledo, and Santiago de Compostela."[44] While he was probably right about Granada and Toledo, it seems unlikely that was the case with Santiago. Starting in the midnineteenth century, with the success of Thomas Cook's modern tours, there was a proliferation of tea-table travel books full of engravings and photogravures. In those books dedicated to Spain (Gustave Doré's, for example, published in New York in 1876; or Alexander Wagner's, published in Berlin in 1880), Santiago is omitted, while Barcelona makes an appearance with dramatic views of the cathedral and the old Inquisition building. At the turn of the century, the influential Baedekers, Blue Guides, and Guides Diamant, devote just a few paragraphs to Santiago while Barcelona's merits are explained in some detail for at least ten or twelve pages, edition after edition.[45] Perhaps none are as representative as the French Guides Diamant, whose meager information on Santiago is limited to the cathedral, while the long pages devoted to Barcelona pay less attention to monuments than to streets, allude less to its past than to its future, emphasize not piety and holiness but rather mundane pleasures. These contrasting views of Santiago and Barcelona are not the exclusive domain of tourist guides. Sexologist and anthropologist Havelock Ellis, a contemporary of the Generation of 98, published in 1908 his prominent *The Soul of Spain,* which included just a passing reference to Santiago de Compostela, precisely in comparison with Monserrat: "Monserrat [is] a shrine of almost unparalleled magnificence, although from its natural position it always preserved a certain aristocratic aloofness, and never enjoyed the immense popular fame throughout Christendom of the other great Spanish shrine, that of St. James of Compostela."[46] Other foreign intellectuals who toured Spain and were closer in age to the Generation of 27, such as John Dos Passos and Waldo Frank (*Rosinante to the Road Again* came out in 1922, *Virgin Spain* in 1926), never devote a single reflection, much less a full exposition, to Santiago de Compostela, but they ponder at length Barcelona's precedence in the country's life, particularly its political life, above all other Spanish cities. Latin American travelers like Darío, Ugarte, Gómez Carrillo, and Nervo did not find any traces of modernity in Spain but those immediately discernible in certain cities of the coastal periphery and, above all, in Barcelona.[47]

Catalonia, in fact, is not so perfectly absent from Lorca's book. Furthermore, today's reader of *Impresiones y paisajes,* aware of Lorca's later developments, will find these early, dismissive references to Catalonia—more æsthetic than political—particularly disturbing. On the one hand, they appear to plainly represent some of the more conservative, nationalistic dictates of the Generation of 98; on the other, they clearly misrepresent the sensibility of a traveler, who only a few years later would fall in love with Barcelona, declare himself an ardent "catalanista," and find in the modern city the spatial sense of belonging he never mastered in old Castile: "Barcelona, however, is another thing. Right? There is the Mediterranean, the spirit, the adventure, the high dream of perfect love. There are palm trees, people from every country, surprising commercial announcements, gothic towers and a rich urban flood tide made by the typewriter. I feel so good there, with that air and passion!"[48]

However, during the 1917 trip, the young traveler made Catalonia responsible for just about every new object he found in bad taste along the way. For example, one of Lorca's complaints about Silos is "that it has as an altarpiece a very large print with a plump Catalonian Heart of Jesus, blond and handsome, showing off this brand new flashy hairstyle, and his beard, recently combed by the barber" (*I,* 88). More significant in view of Lorca's future development is the reference to Catalonian cement: "They will destroy the old charm, and will substitute it with buildings of Catalonian cement" (*I,* 131). As it turned out, Dalí, the friend who started initiating Lorca in Catalan culture in 1922, proposed in 1928 "the destruction of Barcelona's Gothic quarter and that in its place would be raised clear and jovial architectures of reinforced concrete. Propagate the idea that cement really exists."[49] The last reference to Catalonia in Lorca's book is a bit more cryptic but not less meaningful: Lorca uses the word "urbanize (horrific dirty word)" to denounce the projects of Baeza's mayor (*I,* 131). The word was of recent coinage, a creation of Ildefons Cerdà—the engineer who designed Barcelona's urban expansion or "Eixample," "the great nineteenth-century monument to rationality and grid-like order" and, possibly, the referent of Lorca's own "urban flood tide" of later years. In fact, such a concept (and affirmation) of urbanization was part of a larger Catalan nationalist project which did achieve an international unification, well beyond the limited national unity that the *casticismo* [traditionalism] of the Generation of 98 tried to inculcate in Lorca's generation: "In many respects, the Catalan nationalist project is inseparable from universalist, cosmopolitan, and internationalist projects, from the drive to a world market and the prospect of a reunited humanity beyond the narcissism of small differences."[50]

Notes

All translations from the Spanish are by Alice Poust and María López unless otherwise indicated.

1. Luis García Montero, "El taller juvenil," in Andrés Soria Olmedo, ed., *La mirada joven: Estudios sobre la literatura juvenil de Federico García Lorca* (Granada: Prensas Universitarias, 1997), 72 and 84.

2. Miguel de Unamuno, *Por tierras de Portugal y de España* (1911), *Obras completas*, vol. 1 (Madrid: Escélicer, 1966), 281. Unamuno's comment appears in the context of a trip "across Avila's lands." For him, Avila is the most representative Castilian town: "And in Castile there are few cities, if any, superior to Avila" (275). In *Impresiones y paisajes* (see n. 3, below) Lorca characteristically concurs: "Avila is the most Castilian and august city of all the colossal plateau" (24).

3. All quotations from Lorca's *Impresiones y paisajes* refer to the facsimile reproduction of the first edition (Granada: Editorial Don Quijote), 1981, 21. Hereafter *I*, cited in the text.

4. Azorín, *El paisaje de España visto por los españoles* (Madrid: Renacimiento, 1917), 31.

5. This is not always evident throughout the text, but on the very last page the book is dedicated, somewhat arrogantly, to the teacher and classmates "who accompanied me on my trips." The classmates complained: "It is not that we accompanied Federico on his trips or anything of the sort. It is that all of us, including Federico, accompanied Don Martín." See Ian Gibson, *Federico García Lorca*, vol. 1: *De Fuente Vaqueros a Nueva York* (Barcelona: Grijalbo, 1985), 185.

6. Ibid., 164 and 177.

7. I have explored this divergence in "Cartografías del 98: fin de siglo, identidad nacional y diálogo con América," *Anales de la Literatura Española Contemporánea*, no. 23 (1998), 117–45.

8. Federico García Lorca, *Epistolario completo*, ed. Andrew Anderson and Christopher Maurer (Madrid: Cátedra, 1997), 35, 38, and 41.

9. Unamuno had written in 1912: "The Peninsula is healthier than that terrible Spanish America, pantheon of vanities and monstrous envies. At the site of the Burgos cathedral, the city walls in Avila, the aqueduct of Segovia, many miseries of the spirit disappear." See his *Epistolario americano (1890–1936)*, ed. Laureano Robles (Salamanca: Ediciones Universidad de Salamanca, 1996), 396.

10. Darío de Regoyos (1857–1913), so dear a painter to the Generation of 98, and duly praised in *Impresiones y paisajes* (105), painted Burgos's cathedral first in 1900, at night, with the moon behind the towers, against an otherwise purple sky, as if trying to represent both Ugarte and Lorca's view on a single canvas.

11. Manuel Ugarte, *Visiones de España (Apuntes de un viajero argentino)* (Valencia: Prometeo, n.d.), 37–38.

12. Miguel de Unamuno, *Obras completas*, vol. 6 (Madrid: Escélicer, 1968), 999. In *Impresiones y paisajes*, Lorca mentions "todo un espanto Rubeniano [*sic*] hacia la muerte" [a real Rubenian fright of death] (119), an obvious reference to "Lo fatal" and the poems Unamuno preferred. However, Lorca does not dismiss Darío's "sensualidad acústica" [acoustic sensuality]. Quite to the contrary: in an article published in *El Diario de Burgos* and left out of *Impresiones y paisajes* he would precisely underscore the exemplarity of one of Darío's *guitarradas*: "Marcha triunfal" [triumphal march]. Quoted in Gibson, *Lorca*, 1:173.

13. Gibson, *Lorca*, 1:120. Gibson had pointed out earlier: "During that phase in his life, Federico was a pianist above anything else" (108). About music's general representative value—"certain musicality went beyond the purely auditive"—in Lorca's early writings, see Christopher Maurer's introduction to Federico García Lorca, *Prosa inédita de juventud*, ed. Christopher Maurer (Madrid: Cátedra, 1994), 16–21.

14. Generally in the same form of synaesthesia favored by both *Parnassiens* and *modernistas*, as in "El sol se oculta y del monte nacen cascadas infinitas de colores musicales que se precipitan aterciopeladamente sobre la ciudad y la sierra" [The sun hides and the hill gives birth to endless cascades of musical colors that velvetously precipitate over the city and the mountain range] (*Impresiones*, 169). The use or, rather, abuse of the term "acorde" is particularly marked: from "el último acorde de luz" [the last chord of light] (107) to "el acorde dorado de la catedral" [the golden chord of the cathedral] (158) to "acordes de verdor lujurioso y humos sucios" [chords of lustful greenery and dirty smoke] (171), and beyond. Predictably, chapter 4, "Sonidos," in the "Granada" section of the book, is especially crowded with musical references, among them this totalizing proposition: "Podría decirse que suenan todas las cosas . . . Que suena la luz, que suena el color, que suenan las formas" [It could be said that all things sound . . . that light sounds, color sounds, forms sound] (166).

15. This is how Donald Jay Grout characterizes Handel in his famous *A History of Western Music* (New York, Norton, 1973), 436. Grout continues: "his music has German seriousness, Italian suavity, and French grandeur. These qualities were matured in England, the soil then most favorable to the development of such a cosmopolitan style." One could hardly imagine a combination further removed from the idea of the genuine national character propounded by the Generation of 98.

16. If I may resort again to Unamuno as the paradigmatic traveler of that generation, he wrote in 1906, regarding Zamora (one of the towns Lorca visited with Berrueta): "The beauty of the city and its historic sites, like that of the landscape, is, in great part, an effect of literary suggestion. Writers, and especially poets, have taught the people to see the beauty of their own habitats" (*Obras completas*, 1:627). That is precisely the operating principle behind Azorín's *El paisaje de España*, where he writes, for example: "Castile, our Castile has been made by literature" (67).

17. An extraliterary report seems to support this hypothesis. Fourteen years after the trip, Lorca told a Cuban friend about his visit with Unamuno in Salamanca: they spoke about music, "y el chicuelo (Lorca) se levantaba como una afirmación contra la sabiduría del viejo a quien nadie nunca osaba discutir. Federico, no: Federico le negaba muchas cosas: ¡claro! Porque Federico sabía música y don Miguel de Unamuno no sabía" [and the kid (Lorca) stood his ground as an affirmation against the wisdom of the old man, with whom no one dared to argue. Federico, no: Federico said no to him about many things: of course! Because Federico knew music and Don Miguel de Unamuno did not]. Quoted in Anderson and Maurer, eds., García Lorca, *Epistolario completo*, 32, n.30.

18. Miguel de Unamuno, *En torno al casticismo* (1895), ed. and intro. Jon Juaristi (Madrid: Biblioteca Nueva, 1996), 115. Unamuno understood that such a concept of "sensuality" (which also equaled the "sensual" and the feminine) defined best the French landscape and lifestyle as well as the literary style of Latin American modernists: "We tend to note the American style as weak, "madamized," lacking vigor and robustness" (*Epistolario americano*, 147).

19. Of course, like gardens and ruins, lustful monstrosities, however repugnant, are also (perhaps primarily) a pretext for yet another literary exercise: "Podía ser animal raro o hermafrodita satánico. Carne sin alma o medusa dantesca. Ensueño de Goya o visión de San Juan" [It could be a rare animal or a satanic hermaphrodite. Flesh without soul or a Dantesque medusa. Goya's daydream or Saint John's vision] (*Impresiones*, 162). Moreover, what appears to be here a new, uncharted territory, a challenge to the young writer, was in fact a well-established, already conventional topic even in certain Spanish literary circles, particularly

around a minor novelist called Antonio de Hoyos y Vinent (1885–1940), who in 1903 started publishing narrative works with telling titles such as *Del huerto del pecado, La vejez de Heliogábalo, Obscenidad, El caso clínico, Sangre sobre el barro (paisajes patológicos)*. In this regard, perhaps there is nothing as indicative of Lorca's position as his frequent but irresolute use of such a loaded term as "lujuria" [lust]: from the extremely positive—"¡Bien merecía la tarde, cargada de lujurias celestes, un beso apasionado de aquellos amantes!" [The afternoon, full of celestial lustfullness, deserved well a passionate kiss from those lovers!] (238)—to the extremely negative—"La figura monstruosa . . . salía de aquella casa de vicios espantosos y lujurias extremas" [The monstrous figure emerged from that house of frightening vices and extreme lewdness] (162).

20. Gibson, *Lorca*, 1:168–71. Eutimio Martín, in *Federico García Lorca, heterodoxo y mártir. Análisis y proyección de la obra juvenil inédita* (Barcelona: Siglo XXI, 1986), 154, relates Lorca's critique to his reading of San Juan Clímaco's *La escala espiritual*. Neither Gibson nor Martín quote two sentences that, apropos the ornamentation of old tombs, seem to summarize Lorca's somewhat apocalyptic as well as (religiously) simplistic position toward sexuality and flesh: "nadie, absolutamente nadie se librará de los pecados que tanta miel y tanta amargura tienen" [nobody, absolutely nobody will be free of the sins that contain so much honey and bitterness in them] (*Impresiones*, 118–19); "la carne es en la vida lo que manda, dejemos pues que en la muerte viva el alma" [the flesh rules in life, so then lets leave the soul live in death] (120–21).

21. Moreover, this is a reiterated, if not altogether constant, proposition of the traveler. He wrote about mass at Silos: "Son los monjes que ofician hombres de tez curtida, de andar grosero, de manos impuras por el color negruzco que tienen, llenas de cerdas, ese castigo cruel de la naturaleza. Seguramente el prodigioso altar temblará. Debiera [*sic*] por estética no permitir a estos hombres decir la misa, tocar el cáliz de aristocracias santas, alzar la hostia sublime símbolo de pureza y de paz universal" [The monks who officiate are men of weather-beaten skin complexion, of gross walk, their hands made impure by their blackened color, filled with bristles, that cruel punishment of nature. Surely the prodigious altar will tremble. By reason of aesthetics, these men should not be permitted to say mass, touch the saintly aristocratic chalice, or raise the host, sublime symbol of purity and universal peace] (*I*, 94).

22. Berrueta used to call Lorca "our Epulon," for being the richest in the group and for his tendency to give too generous tips (Gibson, *Lorca*, 1:172).

23. Unamuno is referring to Castile: "Villagers do not walk through these dry fields, because they do not exist here; local men of the color of the earth arrive, raised up on their mounts" (*Obras completas*, 1:126).

24. Line 17 from "Por tierras de España": Antonio Machado, *Campos de Castilla (1907–1917)*, ed. Geoffrey Ribbans (Madrid, Cátedra, 1989), 135.

25. Pío Baroja, *Camino de perfección* (1902) (New York: Las Américas, s/d.). The text continues: "caras hoscas por costumbre, gente de mirada siniestra y hablar dulce. En aquellos tipos se comprendía la enorme decadencia de una raza que no guardaba de su antigua energía más que gestos y ademanes, el cascarón de la gallardía y de la fuerza" [frowning faces by custom, people of a sinister look and sweet talk. Those types manifested the decadence of a race that kept only their gestures and attitudes from their former energy, the shell of elegance and strength] (74).

26. Gibson finds "the authentic voice of the born writer" in the piece "Un hospicio de Galicia" and its strong denunciation of "grave social injustices"; see Gibson, *Lorca*, 1:123 and Lorca, *Impresiones*, 248. It seems to me that, on the one hand, that kind of topic and compassionate outcry is rather frequent in the work of his predecessors, from Galdós to Baroja; on the other, Lorca's own voice in this text may not be heard as much in any single fragment as in the general assortment of divergences and dissonances I am trying to analyze here.

27. Or, as he puts it in one of his letters home: "He aprendido muchas cosas y de mucha

necesidad para *una persona educada finamente*" [I have learned many things necessary to *a finely educated person*] (*Epistolario*, 44, my italics).

28. This remark may be worth quoting in full to underscore my point: "Los viajeros eran personas vulgares, con gestos de idiotez, que ansiaban subirse pronto, no les fueran a quitar el sitio, gentes que no veían la maravilla solemne de las lejanías" [The travelers were vulgar people, with idiotic gestures, that anxiously wanted to board quickly, so that their places would not be taken, people that did not see the solemn wonder of the long distances] (*I*, 65).

29. Azorín, *Obras completas*, vol. 1 (Madrid: Aguilar, 1975), 754.

30. Gibson, *Lorca*, 1:104.

31. Unamuno, *Obras completas*, 1:639.

32. In this regard, I find particularly weighty a correction in one of the early manuscripts of the book. Lorca had written: "En una noche negra y lluviosa llegamos a la ciudad de los grandes recuerdos." [On a black and rainy night, we arrived at the city of great memories.] He then crossed out "llegamos" [we arrived] and wrote "llegué" [I arrived] (*Prosa inédita de juventud*, 465). In the writings of the Generation of 98 there is a certain tendency to make a collective experience out of the individual's often solitary trips; Lorca aspired to make a single, hardly transferable experience out of an originally collective one.

33. 1917 was a particularly agitated year in Spanish politics: the trip to Burgos coincided with a general strike, socialist uprisings, and radical censorship of the press. Such consequential public circumstances left no perceptible trace in Lorca's book, but they did in at least one of his letters: "Tengo escritos unos artículos que se publicarán en Madrid cuando pase la marejada." [I have written some articles that will be published in Madrid when the commotion passes] (*Epistolario completo*, 44). Lorca's unelaborated letters home provide a wealth of concrete information that is patently absent from the more carefully crafted book. Did he consider his family less informed and cultivated than the implied reader of his publication, or were his private letters simply more submissive to the journalistic standards of the masters than the public text? The constant, deferential references to Berrueta in the letters, in contrast to the book, appear to support this second hypothesis.

34. Quoted by Antonina Rodrigo, *García Lorca, amigo de Cataluña* (Barcelona: EDHASA, 1984), 219.

35. Quoted by Gibson, *Lorca*, 1:165.

36. Unamuno, *En torno al casticismo* (1902), in *Obras completas*, 1:795. Azorín, *Las confesiones de un pequeño filósofo* (1904), in *Obras completas*, 1:642.

37. Unamuno, *Obras completas*, 1:802.

38. Edward W. Soja, *Postmodern Geographies: The Reassertion of Space in Critical Social Theory* (London: Verso Books, 1989), 24.

39. From "la emoción de lo pasado" [the emotion of the past] in the old Castilian towns to an apostrophe like "¿Qué os dirán las generaciones venideras? ¿Qué saludo os hará la aurora sublime del porvenir?" [What will upcoming generations say to you? What salutation will the sublime dawn of the future give you?] (*Impresiones*, 87 and 19–20).

40. Unamuno, *Obras completas*, 1:637 and 69.

41. Azorín, *Los pueblos*, in *Obras completas*, 1:705. Azorín continues: "Do not envy the wide, symmetrical and worldly routes of the great universal capitals; do not listen to the modern and terrible architects." This may be a covert reference to Barcelona, part of a general disposition I will try to explain shortly.

42. Obviously to be compared with the general "silencio" [silence] Azorín found in Granada: "Se gozaba el silencio" (*El paisaje de España*, 117).

43. The reader may propose that the best choice would have been the inclusion of both places in the itinerary, thus creating a well-balanced axis East/West, Old/New, Sacred/Profane, Time/Space. One wonders, though, whether the geographic ideology of the Generation of 98 was up to this liberal combination.

44. Unamuno, *Obras completas,* 1:627.

45. Of course, neither Santiago, nor even Barcelona, could possibly compete with the detailed attention these tourist guides devoted to Granada (and, in general, to all the allegedly *oriental* towns in Spain, below the fortieth parallel).

46. Havelock Ellis, *The Soul of Spain* (Boston: Houghton Mifflin, 1937), 373.

47. See, for example, one of Darío's comments on Barcelona: "Outside the permanent, the historic, a modern wind triumphs here, bringing something of the future. Of course, I already know that in Madrid, I will find myself in another atmosphere, for if a resounding French style exists here, it has entered through a window open to universal light, an openness that, without a doubt, is better than locking oneself inside four walls and living off the scent of old things." See Rubén Darío, *España contemporánea* (1901), in *Obras completas,* vol. 3 (Madrid: Aguado, 1950), 39. Or Ugarte's "Let others praise contemplative towns that become stagnant and mummified in a fakir pose; we candidly pronounce our admiration for the industrial districts that increase the common welfare, that facilitate life and spread progress" (*Visiones de España,* 164).

48. From a letter to Melchor Fernández Almagro, quoted by Rodrigo, *García Lorca,* 38; Rodrigo also quotes Carles Sindreu's claim that Lorca was "ours in body and soul . . . in this Catalonia which he claimed as his second homeland" (348).

49. Rodrigo, *García Lorca,* 197–98.

50. I take both the information about "urbanización" [urbanization] and the ensuing quotations from Brad Epps's as yet unpublished manuscript "Modern Spaces: Building Barcelona" (to appear in a collection of essays on the city, edited by Joan Ramon Resina), 1, 5, and 33.

"Tópicos sublimados": Lorca's Female Iconographies in "Eros con bastón"

Candelas Gala
Wake Forest University

"Eros con bastón" (Eros with a Cane) is the eighth of eleven sections in Lorca's *Canciones 1921–1924*, a book which reflects a most relevant period in the poet's life and artistic career, spent mostly at the *Residencia de Estudiantes*.[1] "Eros with a Cane" is particularly interesting for two reasons: it is a masterful display of Lorca's use of wit and parodic humor—the Andalusian *guasa* [joking]—as the rhetorical device for a sophisticated game of intertextuality with entrenched attitudes and well-known female prototypes in Spanish popular iconography; and it is one of Lorca's most ingenious articulations of new aesthetic concerns occupying plastic artists and poets of the twenties. In that sense, "Eros with a Cane" exemplifies the search of the new art for iconicity devoid of the anecdotal, for the predominance of the visual in the line of Gongorine aesthetics, and for the modernist technique of distancing as the means to deal with personal and artistic issues of profound import for the poet.

María Moliner's dictionary describes *guasa* as the irony, sarcasm, or mockery in a statement. *Estar de guasa* implies a desire to joke, not to take the topic at hand seriously, and *guasearse* describes a type of joking devoid of acrimony. Written in the atmosphere of the Madrid *Residencia de Estudiantes*, *Songs* is filled with humorist allusions to private jokes among friends. The dedications of sections and poems of the book to various friends or children of friends echo Buñuel's memories of the *Residencia* as a time filled with a mixture of drinking bouts, brothels, jazz, pranks, and long conversations about profound aesthetic issues.[2] More than in any other book, Ian Gibson finds in *Songs* the "ludic spirit" characteristic of the *residentes* and the expression of "the cheerful, playful, and mocking aspect" of Lorca's personality. Gibson goes even further by placing the artistic value of this book in its mocking tone since it allows Lorca to deal with very serious concerns—unfulfillment in love, the fear of death, the passage of time—with an expression that is always "admiringly restrained, refined and ironic."[3]

Taking parodic *guasa* as the overriding rhetorical device, I intend to explore Lorca's personal and artistic concerns at the time he wrote these poems, and to elucidate his position vis-à-vis artistic trends and movements of the times as well as beliefs and expectations of his society regarding sexual preferences and behavior. As Lorca's critical studies move away from the commonplace assessment of his work as the embodiment of Spanishness par excellence and more toward determining his placement within a European frame of reference, the study of writings coming from a period of rich contacts with a variety of artists and artistic trends, such as "Eros with a Cane," is of paramount importance.

Songs is the expression of a period filled with artistic challenges. Like his peers at the *Residencia,* Lorca was well aware of the surrounding artistic world in Madrid and beyond. Avant-garde experimentation, the Gongorine aesthetics of purity, and the surrealist explorations into Freud's subconscious were all part of Lorca's experience at the time. Friends like Dalí and Buñuel had traveled to or were living in Paris, just as Spanish artists of the stature of Picasso, Miró, and Gris had been drawn to the French capital by the possibilities it offered. *Songs* captures a time of choices for a poet like Lorca. Artistically and personally, he must have felt pulled by multiple and often contradictory paths. His love and knowledge of Spanish traditional poetry was coming face-to-face with new experimentations with language; consequently, his deeply rooted love for his country must have been challenged by his equally strong rejection of entrenched Spanish attitudes that left no room for personal expression and difference.

"Eros with a Cane" illustrates Lorca's crossroads very eloquently. The seven poems in this section offer a sort of portrait gallery of female prototypes well known in the Spanish (and sometimes universal) popular iconography. Their recognizable features acquire a fresh look under Lorca's purely humorous, biting, sympathetic, or critical parodic *guasa.* Each is also an example of Lorca's application of Gongorine aesthetics of the gaze which he elaborated in the lecture "La imagen poética de don Luis de Góngora" (The poetic image of Don Luis de Góngora) delivered in February of 1926. Because of their objectivity and attention to the image, Gongorine aesthetics were ideally suited to contemporary artistic concerns of distancing and purity. By choosing objective correlatives within very

Spanish human prototypes and attitudes, these techniques of distancing and parodic intertextuality offered Lorca ways to deal with the rigidity of Spanish attitudes and values and to challenge their open opposition to any type of divergence.

The depiction of Eros with a cane evokes two contradictory images—the self-assured dandy and the hesitant Eros in need of a cane to walk. As the dandy, this Eros embodies the figure of the *flâneur* [dawdler], the epitome of the subject in modernity that Walter Benjamin describes wandering the city streets, looking about, particularly at women. In this sense, "Eros with a Cane" presents the scrutinizing and parodic gaze of a dandy, a *flâneur* reviewing a gallery of seven female stereotypes. However, the *flâneur*'s parodic gaze is also turned on Eros, exposing him as the inadequate male counterpart of the portrayed women. Both meanings remain superimposed in a kind of dialectical exchange that is never fully resolved: Eros is both satisfaction and failure, assurance and insecurity. This ambivalence permeates the poems endowing them with the parodic *guasa* as their most characteristic feature.[4]

Parody is effective in the intertextual play between two or more texts, because according to Linda Hutcheon, and in Bakhtinian terms, it is a "two-voiced" discourse, "an intentional dialogized hybrid." The interplay between "the parodic foreground and parodied background" becomes central since "one text is set against another with the intent of mocking it or making it ludicrous." The distance between the two texts is determined by irony, which, in parody, "comes not from humor in particular but from the degree of engagement of the reader in the intertextual bouncing between . . . complicity and distance." It therefore requires the participation of the reader, who will be called to decode the two or more codes intersecting within the poematic space. The reader must have the capacity to perceive the interplay between the two texts and to decode "texts in the light of other texts."[5] Lorca's parodic *guasa* implies the reader's ability to catch on, to respond to his eye-winking. By dedicating his poems to friends, Lorca is sending messages filled with complicitous humor.

Moreover, because of its focus on self-referentiality, self-reflection, and intertextuality, parody is a common feature of modernism and avant-garde literatures, as Hutcheon notes.[6] It is therefore a technique that flourishes in periods of cultural sophistication such as the one surrounding the writing of *Songs.* Most critics consider *Songs* to be Lorca's most avant-garde and European book.[7] As reflections on art and artistic creation, Lorca's poems exemplify the artistic awareness of modernism. In that sense, they are not only parodies of stereotypical attitudes about sex and of the speaker's own shortcomings with his desire, but are parodies of artistic anxiety as well. Through his gallery of female prototypes, Lorca's speaker parodies entrenched attitudes about sex, desire, and artistic accomplishment. In that sense Lorca achieves what Hutcheon calls a "transcontextualization," a repetition of stereotypes while adding to them the difference brought about by parody.[8]

Christopher Maurer is correct in asserting that "the poetic idea that underlies . . . *Suites*" (and I will add *Canciones* [Songs] as well) "is that a phenomenon . . . can be captured better in a series of random 'momentos' [moments], 'engravings,' or 'vistas' [views] than in the sort of discursive, declamatory verse that one finds throughout his first book, *Libro de poemas*."[9] Each of the seven poems in "Eros" fits the description of "vista" or "engraving" since it captures the essential features of the portrayed female type. These iconographies respond to the goals of cubist artists, as described by Cassou, to break away from discursiveness in favor of the discontinuous, the fragmentary, the instantaneous vision.[10] They respond to the avant-garde goal of achieving "plastic metaphors" as transpositions of real objects into the poetic space. They partake of some of the elements found in ekphrastic poetry because they provide verbal renditions of prototypes whose iconographic value is easily recognized. They are endowed with "enargeia," the rhetorical figure which refers to the capacity of language to provide a description of the object so vivid as to recreate it in the reader's imagination. While they present this seemingly realistic bent, these poems also add an intense awareness of their artificial nature and a highly self-reflexive quality. In this way, the poems of "Eros with a Cane" fuse realism with artificiality, stasis with change, tradition with artistic innovation.

Lorca's interest in the pictorial and the visual is apparent from his many drawings. It is not surprising then that he would be attracted to the visual quality of Gongorine aesthetics.[11] "It is thus that all images open up in the field of vision," writes Lorca in his lecture "La imagen poética de don Luis de Góngora." Lorca's generation shared Góngora's desire to create a new aesthetics, "a new modeling of language," as Lorca described Góngora's art.[12] Like his friends, Lorca was attracted to the objectivity of Góngora's style of writing, to the architectural nature of his images, and to the sensorial quality of his verse.[13] As Nadine Ly reminds us, "what Lorca most appreciated in Góngora was 'the consciousness of his aesthetics.' "[14] Lorca's pictorial approach to his female models draws from Góngora's focus on the signifier at the expense of the signified.[15] The result is not so much the faithfully realistic rendition of the object, but of the impression the object causes in the observer.

Lorca's concern with sight is also an intrinsic part of the artistic project of modernism. In it, the perceptual experience takes over the rational in an attempt to explore language and its relation with reality. Language achieves iconicity, and metaphor is highly valued for its ability to provide new perspectives on the world by connecting disparate elements. For Timothy Reiss, the importance of the visual perception of reality in modernist aesthetics is represented by Galileo's telescope which, as language, places itself between the perceiving eye and the object being perceived.[16] This is the relationship Lorca's poems explore. As they present the speaker's gaze over the female body, the incongruence and distance between his desire and the body itself is made evident. The poem is the space where that incongruence is displayed with all its parodic elements.

Each of the females in the verbal portraits is depicted metonymically. Like in Lorca's drawings from this period, these verbal portraits aim at capturing the "essential characteristics" of the person or object.[17] Their crisp and bold character and their sculptural quality are common to the postwar photographic style which has been attributed to modernism.[18] Lorca's verbal portraits of women are far removed from the sentimental depictions of women with vague contours and ethereal bodies of the pictorialist period. Like in photography from the period by artists such as Imogen Cunningham and Edward Weston, Lorca's gaze is devoid of any mythic and allegoric intention as it stalks the female body "as a subject in itself."[19] The overriding parodic demythification that Lorca carries out in these poems centers mainly on the accepted belief of Eros as fulfillment and of the woman as muse for the male artist. In Lorca's treatment, the woman remains the passive conventional object under the male speaker's gaze, a body predefined by long-standing expectations. By the same token, the passivity and predictability of these female stereotypes become a challenge to the Eros/speaker, mirror-surfaces where his own prowess as male and artist comes into question. Parodic *guasa* is Lorca's way to lighten the topics at hand, thus responding to Harold Bloom's depiction of parody as a way for modern authors to deal with the "anxiety of influence."[20] I would add that it is Lorca's way to deal with personal anxieties about self and sexuality, as well.[21]

"Susto en el comedor" (Fright in the dining room) refers to an episode that could have happened in the dining room of the *Residencia*.[22] The brevity of the lines requires full involvement of the reader in the task of deciphering their allusiveness. Once the scene is established—the woman's fright at thinking that the man was reaching for her breasts—the woman becomes the target of the parodic *guasa,* since the man rejects her in favor of the apples on the table. By making fun of the woman's expectations, the speaker is parodying the entrenched belief that men are almost always going to react sexually to a woman. The parody ultimately extends to the primordial scene with Eve, Adam, and the apple. Contrary to Adam, who was unable to resist Eve's offer, the speaker asserts his difference by choosing the "real" green apple over the woman's "rose" apples of her breasts.

Many critics have seen in this poem an assertion of homosexuality. It is clear that Lorca is contesting the assumption that males have to be attracted to females, and, in that sense, contesting the centrality of heterosexuality. His choice of the "green apple" would suggest a kind of ambivalent sexuality echoed later in the "Verde, que te quiero verde" (Green, oh how I love you green) of "Romance sonámbulo" (Sleepwalking ballad). On the woman's part, he is also making fun of a culture where repression deprives women of their natural sexuality turning them into prudes. Transposed to the artistic realm, the absence of natural sexuality corresponds to the artist's experience of language and its relation to the world. Contrary to the allegorical and symbolical baggage surrounding Eve's rosy apple, the green one represents the "real" apple devoid of predetermined connotations, the language identical to the object. Therefore, the poem's parodic intertextuality with the Adam and Eve scene refers the reader back to a primordial time when word and object were one. The male speaker, like Adam, sought the real apple, or pretended to view the fruit in its pristine reality. Like Eve, the woman, however, knew that the apple was already charged with further meanings tied to the knowledge of good and evil. By rejecting the woman's apple, or breasts, the speaker does more than simply rejecting heterosexuality as the sexual orientation favored by Western civilization. He is also rejecting a language that has lost its pristine nature by the accumulation of metaphorical meanings of its words. It could also be said that he rejects the woman's fruit, or encumbered language, as the culprit for the perversion of Adam's pristine world, of his sexuality of sameness. The woman embodies a false sense of morality where natural instincts and desires are repressed and chastised.[23]

In a very whimsical, personal style seemingly devoid of seriousness, and through the distancing mode of parody, Lorca articulates very personal and artistic concerns. He shares with modernism the desire for an iconic language freed from the moral and religious baggage placed upon it by tradition and culture. Just as the woman's face in the poem turned from *rosa* [rose color] to *alimonada* [lemon color] upon realizing that she had read too much into the man's intentions, language turns yellow and withers when it departs too much from the world. Lorca will deal with the same issue in later works such as *Doña Rosita, la soltera* (Doña Rosita, the spinster). Moreover, the speaker's artistic self-awareness is evident in the fact that he intentionally tricked the woman. Even though he knew all along how she was reading his gesture, he still led her on.

This reading seems to be confirmed by the closing parenthetical image: "(Grulla dormida la tarde, / puso en tierra la otra pata)" [(The afternoon, a crane asleep, / put its other foot down)], suggesting that the evening as *grulla* has turned into night. The image recalls something Lorca wrote in a letter to art critic Sebastià Gasch: "mis ojos y mis palabras están en otro sitio. Están en la inmensa biblioteca que no ha leído nadie, en un aire fresquísimo, país donde las cosas bailan con un solo pie" [my eyes and my words are elsewhere. They are in the immense library that no one has read, in the freshest air, in a country where things dance on just one foot].[24] The letter alludes to a fresh, pristine world where solid grounding is unnecessary because meanings have not been set and new possibilities are always open. The situation in the dining room is full of expectation, at least until the speaker's intention is made clear. This expectation relies on preconceived notions, ones that the male speaker consciously exploits and thus exposes. By returning to the "apple" its original meaning, the speaker fooled conventional expectations and momentarily returned to that *otro sitio* still fresh from superimposed meanings. The evening/crane image frames the scene in a vignettelike fashion; it thus successfully effects the parodic repetition with a twist. It is also Lorca's objective correlative as it distances the personal meaning of the scene from the poet.

While the speaker in "Fright" rejects the type of poetry the "rose-colored" woman offers him, in "Lucía Martínez" ("EC," 478–79), he openly declares his desire to possess the woman. Under the male's gaze, Lucía Martínez is a body in full view and, particularly, a close-up of the female genitals. The blatant nature of the scene is counterbalanced by an accumulation of stereotypical images leaving no doubt of Lorca's parodic intention. If the woman in "Fright" fit the model of the prude, Lucía Martínez embodies the stereotype of the sensuous female body that Ewing describes as "the nude female supine on a couch, arranged so as to maximize the curve of breast and thigh; the eyes closed or averted, feigning abandon or sleep; the rich, billowing fabric hinting at pleasure of the boudoir" in a décor including "flowing draperies, oriental rugs, musical instruments."[25] While the mention of her first and last names endow her with a sense of pedestrian reality, her metaphorization as "Umbría de seda roja" [Shadowy in red silk] and the series of flower images taken from the pictorialist and symbolist traditions of the portrait, counterbalance that first impression. Lorca could be depicting a male's fantasy about the most familiar woman he may know or encounter. In that sense, the fusion of "everyday life"—or the realness and commonality of this woman—with the process of her metaphorization, touch upon a crucial concern in these poems, that is, the potential to regain the pristine quality of experience and art vis-à-vis tradition and conventions.

The necessary counterpart to this female—sensorial fusion of sight and touch ("Shadowy in red silk")—is the macho speaker who, in the presence of such a woman, will necessarily feel compelled to possess her. Fittingly, then, in direct opposition to the speaker in "Fright," the speaker here declares his intention to consume Lucía's mouth. The determination in his tone, "Aquí estoy, Lucía Martínez" [Here I am, Lucía Martínez], echoes the daring attitude of the bullfighter in front of the bull: the speaker will consume the woman because that is what he wants and can do. Although his pronouncement announces his intention to perform an unmediated sexual exchange, the scene is charged with irony. The parodic *guasa* is aimed here at the macho type who, in view of a woman's body, is forced to desire it because of his declared virile status. However, Lorca makes fun of the macho's unquestioned sexual prowess by resorting to the same Spanish proverb the macho uses, "[P]orque quiero y porque puedo" [Because I want to and I can] or "Querer es poder" [Where there's a will there's a way]. His words *declare,* but the actual action remains unconfirmed.

This poem offers a peek and very personal version of Lorca's longstanding concern with human desire. While his friends Dalí and Buñuel were finding their own ways to articulate such a conflict in the surrealist film *Un Chien andalou,* for instance, as well as in paintings and later films, Lorca chooses his own scenario, the most stereotypical one for the fulfillment of desire—the boudoir—and the most stereotypical characters, the sensuous woman in red silk and the macho. Around these stereotypes the poem plays with the polarity of word and action, desire and fulfillment, *querer* and *poder.* The speaker "declares" his desire but the conventionality of the setting devoids his determination of truth. The female genitalia as magnolias, that "azabaches recónditos" [hidden veins of jet] darken, place Lucía Martínez in the long tradition of sensual oriental poetry and pictorial portraiture. When he proclaims that he will drag her by the hair "en madrugada de conchas" [into the seashells of daybreak], we are provided with a parodic representation of Venus riding on a shell-carriage after emerging from the sea foam. But we are also reminded of the scene in Buñuel's *L'Age d'or* (The golden age) where Modot, the male protagonist, drags his female counterpart's dress in a gesture reminiscent of cavemen. Desire, or the *querer,* is heavily couched or conditioned in conventionality. Although the macho proclaims what he plans to do, the directness of his desire is aimed at a highly codified and predictable scenario. Female sensuousness and male prowess lack the spontaneity expected for a fully satisfying sexual encounter because they have already been highly predetermined. The repetitiveness of the scene transforms it into a caricature.

In poetic terms, Lucía Martínez embodies a view of poetry as a discourse whose highly codified patterns impede a direct, spontaneous articulation of desire. Just as the male has to cope with expectations of his sexuality, the poet has to deal with his own anxiety as a result of conventional artistic expectations. Different literary traditions, with *The Thousand and One Arabian Nights* as a prominent example, have developed the language of love and sexuality so specifically that the poet is hard-pressed to find his own voice. With a great deal of Andalusian *guasa,* Lorca finds a way to deal with his own anxiety of influence.

The parody of official religion in "La soltera en misa" (The unmarried woman at mass] ("EC," 478–79) is flagrant. Under the ominous presence of "el Moisés del incienso" [the Moses of the incense], the spinster Virginia, already fated to a life of sexual unfulfillment by her name, dozes off while praying. The Moses refers to the back-and-forth movement of the incense dispenser. Therefore it brings to mind the so-called baby cradle, which this spinster will never have, and the biblical figure of Moses. The ten commandments Moses brought down from the mountain hang over this woman's head with the same back-and-forth movement of the cradle denied to her frustrated motherhood. That weight comes down in the form of the incense masking her senses and sensuality. The picture vividly portrays Marx's dictum that religion is opium. The male speaker's gaze focuses on Virginia's breasts as "negros melones" [black melons] that he urges the woman to offer "al rumor de la misa" [the drone of the mass]. The altar thus becomes the site where Virginia's sexuality is offered in sacrifice to religion. The speaker also urges her not to stir "con ese traje de profunda seda" [in that dress of yours deep in silk], afraid, perhaps, that if she were to move, the body would be awakened to sexual desire. Freeze the frame, the poetic speaker seems to say, lest the contained picture come undone.

As expected, that fear becomes a real threat in the lascivious "ojos de toro" [eyes of bulls] looking the spinster over from some hidden corner in the church. The grotesque nature of the scene is evident in its inconsistency: piety and modesty are placed side by side with lust, and the religious sacrifice acquires overtones of a black mass. No matter what the church rules, desire expresses itself. The enemy is within, for, lurking in the darkness of the church aisles, the bull of desire is stalking its prey.

Lorca returns to the bullfighting image he suggested in "Lucía Martínez" with the difference that in "The Unmarried Woman at Mass" the speaker is not the protagonist, but the outsider looking in. He observes the church as the arena where desire and repression face one another in opposition. This confrontation of opposing sides echoes that of the poetic speaker face to face with poetry as a discourse constrained by religion and morality. As in earlier poems ("Elegía" [Elegy] in *Libro de poemas,* for instance) and in later works, Lorca chooses a woman to represent religious and moral repression, for she is most directly affected by it. By taking a step back and presenting the scene from the eyes of the speaker/poet, Lorca objectifies the very real effect of repression in his life and writings. As a young poet, he faced all the repressive rules of religion and its demand to silence the flesh. The parodic voice he chooses to articulate that conflict is evidence of greater sophistication and aesthetic distance vis-à-vis his earlier more passionate displays against religion. The church, he seems to be saying, has become a confrontational site for opposing forces rather than the site of reconciliation, peace, and love. Its rules either exacerbate desire or dull it; in either case, desire is distorted, made to be monstrous and unnatural. The breasts of nurturing milk become, in the spinster, the rotten image of black melons, while the multiple pleats of her dress seem to represent the weight of religious formulas and admonitions, the "drone" or "rumor" which, repeated generation after generation, ends by killing vitality, the incense of a religion stifling the "fresh air" of perception. The fight in the arena of the church is one between life and death.

In "Interior" (Interior, "EC," 480–81), from the outset, the speaker proclaims his lack of desire to be a poet or a gallant, both highly codified roles he has played vis-à-vis his female prototypes. In his rejection of these roles, he embraces their antithesis: the uncontrolled annihilation through unbridled sexuality. Carmen, the name of the woman in this poem, recalls the femme fatale of Merimée's novel and Bizet's opera. The speaker calls for "¡Sábanas blancas!" [White sheets!] where she could faint and lose consciousness. And since the bed is the "relumbrante" [dazzling] site where man and woman meet, as the Servant states in *Bodas de sangre* (Blood wedding), the speaker perceives Carmen as the embodiment of the loss of control in an experience of sexual fulfillment. In her total nakedness, Carmen blinds the onlooker and, like the squid, envelops him in the perfume of her sex/ink. The reference to the ink and the white sheets, reminiscent of a white sheet of paper, suggest the act of writing. The speaker would like for Carmen to lose consciousness on his white page, thus dissolving all barriers imposed by reason. He would like to be blinded by her nakedness and black ink in order to obliterate the scene of writing and all of its boundaries. By losing himself in her body of desire, all constraints and distance between desire and object, word and thing, would dissipate.

Why pick the prototype of Carmen to obliterate the roles of poet and gallant that he has played? Why choose her to articulate an unmediated desire that the highly codified discourses of poetry and gallantry do not allow? Carmen is the female Don Juan, the woman men love, despite the tragedy she causes them by not reciprocating their affections. Lorca's parody extends beyond the prototype to the sexual and artistic fulfillment Carmen promises. Just as her counterpart in Merimée and Bizet, her invitation to let go, to lose consciousness, is an invitation to self-annihilation.

Is Lorca parodying his friends, Dalí and Buñuel, and their experimentations with surrealism and the subconscious? Allowing the subconscious to take over would be like blotting with ink the work of art thus dissolving all form. Equally, desire calls for a full surrender, which, at the same time, implies the losing of oneself in the same act where fulfillment was the goal. It is mischievous of Lorca to resort to a prototype which exudes "Spanishness" in order to poke fun at his friends' more international excursions into surrealism, showing that both yield the same dangerous results. It is also interesting to compare Carmen with a squid, thus relating her to the French surrealist image of the fish as inhabitant of realms below the conscious level.[26] Like the squid, Carmen ejects a perfume/ink whose blackness blinds daylight and reason. With his Andalusian *guasa,* Lorca could have been sending his friends the message that the subconscious is already present at home with equally dangerous consequences as its French counterpart.

The answer to the questions posed by "Interior" seems to be expressed in "Nu" ("EC," 480–81). This poem can be read as the reversal of "Interior" and as the open rejection of "foreign" fashions that "Interior" merely suggested. If "Interior" recreated a hidden, private space, as the title suggests, "Nu" offers a scene in the open with a female model that, contrary to Carmen, lacks all power to attract. Its title refers to the nude genre in plastic arts; the fact that it is in French suggests a realm of greater sexual freedom than the one existing in Spain, at least at the time the poem was written. Right from the start Lorca would seem to be criticizing the all too obvious display of the female body because it eliminates the essential ingredient in desire and passion: mystery. This reading is confirmed in the initial couplet where the French naked woman is depicted in an entirely Spanish setting:"Bajo la adelfa sin luna / estabas fea desnuda" [Beneath the moon-dark rosebay / you looked ugly naked]. The Spanish *piropo* [compliment] is here reversed in a direct insult to the French woman and her nakedness.

The rejection of the woman in "Nu" recalls the same sort of rejection in "Fright in the Dining Room." In both cases the speaker makes a conscious point of rejecting the woman's more or less overt offer of sexual favors. If in "Fright" the male speaker parodies the long-standing belief that men will be attracted to women and a false sense of female modesty that obliterates the direct expression of desire, in "Nu" he offers his own version of pure poetry and artistic freedom. While Jiménez wanted his woman/poetry denuded of "false rhetoric," and surrealism called for a release of all inhibitions, Lorca points to the danger of total nakedness or excessive purity and asepsis. This naked woman, like the whiteness he associated with Jiménez in his "portrait with shading," is devoid of all mystery and, as such, is devoid of depth of human emotion.[27] Moreover, the poem addresses the role of other countries, particularly France, in the development of an art that impacted Lorca and his artist friends. The rejection of the French woman here recalls Lorca's nationalistic tone in "Sketch de la nueva pintura" (Sketch on modern painting), where he attributes to Spain the glory of having produced the three great revolutionaries of modern painting: Picasso, Miró, and Gris.[28]

Besides the title, there is further evidence of the role of national boundaries in the second couplet: "Tu carne buscó en mi mapa / el amarillo de España" [Your flesh sought on my map / the yellow of Spain]. The map could refer to his human geography that the line identifies with "el amarillo de España." The first thought that comes to mind is the plains of Castile often identified by that color. Lorca could then be opposing the French woman's naked flesh with the most essential identity of Spain: Castile, with all the connotations of spirituality, mysticism, and honor. The essence of Spain is further developed when "Nu" calls for the "covering" of the body contrary to its "uncovering" or freeing from mediating codes in "Interior." With a great deal of the bullfighter's macho attitude, the speaker of "Nu" covers the French woman's naked body with the green and red cape of his talent. This gesture could mean that the poet offers his talent of life and passion (as symbolized by the green and red colors) to obliterate a type of woman/poetry devoid of beauty, mystery, and passion.

The exposure of the body and the subsequent covering by the cape bring to mind the *taparse/destaparse* [covering/uncovering] polemic in the Spain of the nineteenth century. At that time the Spanish cape went under serious attack from the *afrancesado* [francophile] and progressive forces for what they perceived to be the cape's association with the Spain of the black legend and the Inquisition. Lorca could be reaffirming Spanish identity vis-à-vis the influence of France, as in the nineteenth century. The red and green cape of his talent describes Spanish art with the colors of passion and life, of elementary life that French art, more abstract and rational, did not champion.

The overtly nationalistic final line, "¡Aquí somos otra gente!" [Here we're a different breed!] produces an uncomfortable impression when read from today's perspective. If said in earnest, Lorca is clearly set on establishing the difference of Spanish identity versus foreign influences, and this seems to be the intention of this poem. To attempt an art that combines French ideas with the Spanish tradition—the bitter *adelfa* [rosebay] of the Andalusian soul—is incongruous. As other modernists, Pessoa, for instance, Lorca offers a curious combination of tradition and innovation. While he was fully involved in experimenting with new artistic techniques, he maintained a deep nationalistic attachment to his Spanish origins. Lorca's rejection of the French woman and of the prudish woman in "Fright" is overt and seems to be aimed mainly at the reliance of these female stereotypes on the irresistible attraction of the female body. Moreover, Lorca seems to reject the falsity of their tenets, as they represent the perversion of sexuality by codes of false modesty ("Fright") or by an extreme transparency that obliterates the necessary ingredient of mystery in art and eroticism.

The intratextual references of "Serenata (Homenaje a Lope de Vega)" (Serenade [Homage to Lope de Vega]) (480–83) to other Lorcan texts reveal the importance of this poem's theme in the author's cosmogony. The same *serenata* is sung to Belisa in *Amor de don Perlimplín con Belisa en su jardín* (Love of Don Perlimplín with Belisa in the garden); the traditional motif of the "bath of love" reappears in later works such as *Diván del Tamarit*'s, "Casida de la muchacha dorada" (Divan of the Tamarit, "qasida of the golden girl"),[29] while one of the earlier appearances of the bouquets fading for love is found in *Libro de poemas*. In "Elegía a doña Juana la loca" (Elegy to Doña Juana, the mad princess) from *Libro de poemas*, the princess who was made for "flores, versos y collares de perlas" [flowers, poetry and pearl necklaces] receives instead Death's gift of "rosas marchitas en un ramo" [a bouquet of withered roses]. The same topic reappears in "Casida de los ramos" (Qasida of the branches) from *Diván del Tamarit* where the futility of the happiness of the branches is fully manifested, since Death is waiting for them to fall and break down.[30] The incongruence between what is expected and what one gets is at the core of Lorca's fascination with this motif.

The connection with music, as the title "Serenade" attests, is a frequent occurrence in modernist literature where the links with other arts was highly desired. The word *serenata* (as a combination of "serene" and "evening") describes a song sung at night at the lady's window. Lorca's *serenata* to Lolita stands out for its incongruence, for rather than serenity, the scene is charged with agonizing anxiety, and, while everything around exudes sexual fulfillment, Lolita drowns in her own unfulfilled desire. Lolita's naked body, her breasts and white thighs, stand alone and separated from the surrounding context. In her desolation, Lolita recalls Yerma, who also lamented her sterility vis-à-vis the display of life and fertility in the surrounding fields.

The three stanzas describe a temporal progression and its effect on desire. In the first two, Lolita and the night are placed at opposite ends; while the night exudes sexual fulfillment, Lolita is consumed by her own desire. From wading on the river banks and singing across the bridges, the night moves upward until it reaches the rooftop tiles. This ascension marks the progression of the moonlight from its reflection in the river waters to enveloping the whole scene in its silver light. The moon achieves its triumph in the third and final stanza where the night steals the "Plata de arroyos y espejos" [Silver of streams and mirrors] and the anise of Lolita's white thighs. The moon has played its usual game of seduction, thus fooling Lolita's innocent eroticism. Lolita's final dissolution into the moon's whiteness catches us by surprise. In the game of reflections played by the "Plata de arroyos y espejos," the parodic *burla* [joke] of this poem is made evident. The *serenata* to the lady becomes a lament, since Lolita is obliterated in the deadly whiteness of the moon. The chorus about the bouquets, with its progression from "fading" to "starved" and, finally, to "die," narrates love's tragic story vis-à-vis time and death.

"Serenata" is a disturbing poem because of the incongruity between its carefree appearance and its heavily charged content. Under the guise of serenity, of an apparently simple and transparent poetry in the line of Lope and traditional *cancioneros* [song or poetry collections], Lorca has drawn a picture of cruel mockery. Lorca has created a minidrama to objectify his deep fears concerning frustration, time, and death. Lolita embodies an innocent eroticism that is threatened by time and death from the outset. She is also the embodiment of a type of popular poetry with which Lorca had been well acquainted since childhood.[31] Her desolate body becomes pathetic through Lorca's representation of the impossibility of maintaining faith, hope, and happiness in the face of death. By resorting to a traditional motif and to a form clearly identified with popular poetry, Lorca exhibits the dark side hidden behind that apparently calm exterior. The meaning of this poem lies in the incongruity between the poem's sweet, popular form and its dark content. Parallel to the seemingly unthreatening, lyrical game between stanzas and chorus, the moon is playing tricks on Lolita's unsuspecting desire.

The female protagonist of "En Málaga" (In Malaga) is "Suntuosa Leonarda" [Sumptuous Leonarda], a full-figured woman with "Carne pontifical y traje blanco" [Pontifical flesh and white gown]. The speaker describes her as "Ceres en retórica de mármol" [Ceres in a rhetoric of marble] as she moves about swaying her behind "—concha y loto a la vez—" [—conch shell and lotus at once—]. Lorca's parody seems to be targeted toward the papacy, on one hand, and toward Leonardo da Vinci, as the embodiment of classical art, on the other. Leonarda appears as a mature, sexually satisfied woman in control of her life and body, a madame unafraid to exhibit her wares at the balcony of her house of prostitution, "Villa Leonarda," whose balustrades look out onto the streets and seaport of Malaga. Lorca seems to follow in the footsteps of other surrealists, who played pranks on Leonardo's Mona Lisa by turning her into a madame Pope whose Vatican house, like the Vatican in St. Peter's square, opens its doors to the pilgrims of the world.

The "Negros torsos bañistas" [Black torsos of bathers] darkening the seashore offers a visual image of fragmented bodies, just like the person high above in the balcony would see when gazing at the people congregating below. Body fragmentation, a twentieth-century phenomenon stemming from the expansion of photography after the "instantaneous photography" of the previous century, put an end to the idea "of an immutable perception of nature."[32] According to Ewing, the sense of dislocation from the proliferation of body fragments is in itself "an evocative metaphor for the modern condition."[33] The focus placed on these "black torsos" in Lorca's poem provides a view of the human sea as a dark mass lacking wholeness. Devoid of heads, these torsos depict humanity as lacking a spiritual or transcendental guidance. Leonarda is then the pope's reversal as she represents the papacy of desire that the church represses.[34]

Leonarda is a carnival-like figure, as she embodies the underside of the institution of the papacy and of high art. While the church and high art determine what is moral and beautiful according to the head, Leonarda redefines those categories according to the "culo," or tail, the head's antithesis, with its free-swaying motion and obvious sexuality. Leonarda's backside represents all that is not the head, all that is hidden, that is underneath and physical. Like the poem states, she is Ceres, the mother goddess Lorca is offering as an alternative to the church's patriarchal hierarchy. As Hutcheon indicates, "Parody is a form of imitation . . . characterized by ironic inversion."[35] The swaying movement of her corpulent behind offers a parodic reversal of the vertical, ascending direction the church espouses. Leonarda sways, while Virginia, the spinster, was told to remain still for fear that movement would stir up the desire her black dress contained. The double nature of this behind, "concha y loto," suggests several readings. The *concha,* or shell, is suggestive of Venus and Santiago, two figures present in Lorca's writings from his earliest works. If, as critics have done, we associate the shell with the West and the lotus with the Orient, Leonarda would oscillate between the two cultural worlds being, in that sense, a truly catholic pope. Leonarda-Ceres is the earth mother capable of fusing the verbal with the plastic, the word with the form, the temporal with the spatial.

These two images of the shell and the lotus are each suggestive of the vulva. According to Barbara Walker, the legend of Santiago was first heard in the seventh century when the church adopted the sanctuary of the sea goddess Brigit to commemorate the apostle.[36] The symbol of the goddess was the kteis, or vulva in the shape of a conch. The lotus, also an image of the vulva, is personified by the goddess Padma, or Lotus, associated with the primordial vulva from which everything stems and the goddess door, or entrance, to sexual knowledge.

It is telling that "Eros with a Cane" opens with the parody of the primordial apple of knowledge and ends with Leonarda's behind, the entrance to sexual knowledge as a way to spiritual knowledge. It is also noteworthy that Leonarda is the only woman who is described by her movement, while the rest appear as passive objects under the male gaze. One could surmise that Lorca presents Leonarda as a happy fusion of opposites: spirit/body, East/West, nature/art. Modernist and Gongorine attempts to achieve purity at the expense of life are bound to fail, just as transparent art is a naïve assumption. Ceres, or Nature, keeps oscillating between carnality and marble rhetoric, life and classicism. Contrary to Jiménez's naked poetry, or Mallarmé's pure form, Lorca humorously proposes Leonarda as muse, the goddess Ceres where form and content, art and flesh fuse. While the other stereotypes are irremediably fixed in their roles, Leonarda resists all objectification because she keeps oscillating. More than the other prototypes, she responds to Lorca's understanding of art as a struggle with living forms, a constant negotiation with life so that it will not become fossilized.

In "Eros with a Cane," Lorca offers a gallery of female stereotypes as mirrors to reflect a series of cultural and artistic beliefs and expectations. The prude, the femme fatale, the spinster, the seductress, the ugly type, the innocent girl, the sexually knowledgeable woman, represent stereotyped sexual behaviors within the Western Judeo-Christian cultural and religious tradition. In Lorca's poems, they become parodic embodiments of the woman as muse for the male poet. In this sense, they offer the opportunity to parody entrenched sexual attitudes and artistic expectations.

"Eros with a Cane" is one example among many of Lorca's constant experimentation with form, with the point where tradition and innovation meet and separate, and where art and human experience establish their negotiation. Humor and the iconic are the means to avoid the discursive and the sentimental; the male gaze over a fragmented body puts into question ideas of permanence and wholeness. The exploration undertaken through this gallery points to Lorca's artistic stance vis-à-vis the aesthetics of the time. While his friend Dalí was seeking to control emotion via the aesthetics of "Saint Objectivity," Lorca was articulating what he later termed as the "criterio poético-plástico o plástico-poético en justa unión" [the poetic-plastic or plastic-poetic criterion in exact union].[37] It is possible to surmise that Lorca's choice of the *guasa* tone in "Eros with a Cane" was his way of dealing with differences that would eventually create an irreparable split between him and his close friend, Salvador Dalí.

Notes

All translations from the Spanish are my own unless otherwise indicated.

1. The phrase *tópicos sublimados* in this essay's title comes from one of Lorca's letters to art critic Sebastià Gasch possibly dating from 1927; see Federico García Lorca, *Epistolario,* vol. 2, ed. Christopher Maurer (Madrid: Alianza, 1983), 79.

2. José de la Colina and Tomás Pérez Turrent, *Objects of Desire: Conversations with Luis Buñuel,* ed. and trans. Paul Lenti (New York: Marsilio, 1992), 65.

3. Ian Gibson, *Federico García Lorca,* vol. 1: *De Fuente Vaqueros a Nueva York 1898–1929* (Barcelona: Grijalbo, 1985), 378.

4. The "limping" motif—or some type of difficulty with walking—recurs in a number of Dalí's paintings in the form of crutches, most notably in *Specter of Sex Appeal* (1934), *Autumn Cannibalism* (1936–1937), *The Dream* (1937), *Soft Self-Portrait with Fried Bacon* (1941), and closer to the date of *Songs, Honey Is Sweeter than Blood* (1927) and *Cenicitas* (1928). Buñuel deals with the same motif in his 1970 film *Tristana,* although his feet and leg fixation is present in most, if not all, of his films.

5. Linda Hutcheon, *A Theory of Parody: The Teachings of Twentieth-Century Art Forms* (New York: Methuen, 1985), 69, 22, 31–32, 37.

6. Ibid., particularly 72, 96.

7. Luis Felipe Vivanco, *Introducción a la poesía española contemporánea* (Madrid: Guadarrama, 1957), 2.

8. Hutcheon, *Theory of Parody,* 7–8.

9. Christopher Maurer, "Perspectivas críticas: horizontes infinitos. Two Critical Editions of Lorca's Early Poetry (A Review Article)," *Anales de la Literatura Española Contemporánea* 14 (1989): 227.

10. Cited in Guillermo de la Torre, *Historia de las literaturas de vanguardia,* vol. 1 (Madrid: Guadarrama, 1974), 235.

11. Eugenio Carmona, in his analysis of the connections between painting and poetry in the 1927 poetic generation entitled "Pintura y poesía en la generación del 27 (Seis aproximaciones)," *Cuadernos Hispanoamericanos,* nos. 433–34 (July/August 1986), 107, perceives a search for homogeneity between writing and icon, the poetic and the plastic. In the case of Barradas, a painter with whom Lorca collaborated, there is an avoidance of discursive and narrative elements also found in Lorca's poetry of the 1920s. It is also possible to apply to Lorca what Carmona sees in Barradas paintings, that "the syntax of his works is based on iconic fragmentation." Referring to the classicist expression of Cubism, Carmona points out that painters and poets sought after precise contours and profiles, after the exact expression of reality through which to capture the essential (109). Lorca's verbal portraits of female prototypes included in "Eros with a Cane" possess a "presentativo" character that corresponds with Carmona's characterization of the art of that period. Moreover, Lorca's attention to body parts corresponds to the cubist attempt to eliminate the anecdotal and descriptive in favor of ellipsis and the fragmentary (see de la Torre, *Historia,* 240).

12. Federico García Lorca, "La imagen poética de don Luis de Góngora," *Conferencias,* vol. 1, ed. Christopher Maurer (Madrid: Alianza, 1984), 101, 97.

13. Ibid., 98–99.

14. Nadine Ly, "Lorca y la teoría de la escritura: 'La imagen poética de don Luis de Góngora,' " *Valoración actual de la obra de García Lorca.* Actas del coloquio hispano-francés celebrado en la casa de Velázquez, 13–14 de marzo, 1986 (Madrid: Casa de Velázquez/Universidad Complutense, 1988), 179.

15. In her analysis of Lorca's lecture on Góngora, Ly establishes connections between Rousseau's theory of language, as elaborated by Derrida, and Lorca (ibid., 179). Her use of the terminology of the signifier and signified polarity helps to clarify Lorca's often unclear explanations of Góngora's aesthetics.

16. Timothy J. Reiss, *The Discourse of Modernism* (Ithaca, N.Y.: Cornell University Press, 1982), 25–26.

17. Lorca, *Epistolario,* 2: 79. In that same letter of September 1927 to Gasch, Lorca discusses what he calls his "poetic-plastic" or "plastic-poetic" criterion for art, which the Catalan critic referred to as "Totalismo." For a thorough discussion of Gasch's "totalismo" and his corre-

spondence with Lorca, see Cecelia J. Cavanaugh, *Lorca's Drawings and Poems* (Lewisburg, Penna.: Bucknell University Press, 1995), 36–72.

18. William A. Ewing, *The Body: Photographs of the Human Form* (San Francisco: Chronicle Books, 1994), 27. Hereafter *B*, cited in the text.

19. Ibid.

20. See Hutcheon, *Theory of Parody*, 96.

21. Gustavo Adolfo Bécquer's "Rima XI" and Juan Ramón Jiménez's "Vino primero, pura" come to mind as possible subtexts for Lorca's gallery of female prototypes. Those "master" texts would represent the parodied background against which Lorca's parodic texts emerge and with which they interact textually. Bécquer's poetry, and Romanticism in general, played a crucial role in the early development of Lorcan aesthetics; Lorca's writings from the 1920s are mostly a reaction against romantic and symbolist artistic tenets. Regarding Jiménez, Lorca and he met in 1919 at the *Residencia*. In 1921 Jiménez published in *Indice* three of Lorca's poems and Lorca included in *Songs* a portrait of the older poet "with the shadow of Venus." The importance of Jiménez for the young poets of Lorca's generation is indisputable. Equally indisputable is the problematic position that his leadership—as the embodiment of the ideal of pure poetry—came to represent for many of them. This ambivalence is evident in Lorca's portrait of Jiménez, in which the young poet's admiration for the high ideals Jiménez aimed to reach for his poetry is mixed with a sense that such extreme asepsis would be deadly for art.

According to Juan Carlos Rodríguez, *Lorca y el sentido: Un inconsciente para una historia* (Madrid: Akal, 1994), Jiménez and Lorca share aesthetic concerns about form and language, what the critic refers to as " the problem of expression" (15). Both share the interest in phenomenology of visual perception as a way to capture "la detención del instante en el ojo, como la retención visual o espacial del tiempo" [the arrest of the instant in the eye, like the visual or spatial retention of time]. For Rodríguez, Juan Ramón is more "intentional" from outside of the poem while Lorca is more "intentional" from inside of the poem: "Juan Ramón Jiménez represents *aesthetics turned into ethics*; Lorca represents *ethics turned into aesthetics*" (115).

For critic José Luis Cano, "Nota sobre una fijación infantil de Lorca: los muslos," *Cuadernos Hispanoamericanos*, nos. 433–34 (July/August, 1986), 139, the praising of parts of the female anatomy comes mainly from Arabic-Andalusian and Golden Age poetic traditions. Cano surmises that Lorca's fixation on thighs may come from his childhood.

22. Gibson, *Lorca*, 1: 379, identifies the dining room of "Fright" with the one at the *Residencia*. All references to "Eros with a Cane" come from Federico García Lorca, *Collected Poems: A Bilingual Edition*, ed. Christopher Maurer (New York: Farrar, Straus, Giroux, 1991), 476–83. Hereafter "EC," cited in the text.

23. The name of Adam seems to connote for Lorca a pristine state of being while Eve is associated with decrepitude. In "Poema doble del lago Edén" (Double poem of lake eden) from *Poet in New York*, in *Obras completas*, vol. 1, ed. Arturo del Hoyo (Madrid: Aguilar, 1973), Eve eats ants while Adam "fecunda peces deslumbrados" [fertilizes dazzling fish] (489); in "El rey de Harlem" (The king of Harlem), Lorca writes that it is necessary to kill "a todos los amigos de la manzana" [all the friends of the apple] (460). Regarding the meaning of the lemon and orange, Sharon Handley, "García Lorca: Poet of the 'inmensa minoría'? or Voice of the Andalusian Pueblo?" *Crítica Hispanica* 18, no. 2 (1996), quotes a traditional poem where the meaning of the lemon tree is related to "Fright." There the lemon tree is "senos donde maman / las brisas del mar / naranjal desfallecido / naranjal moribundo. / Limonar. Tu /viste mi amor roto / por el hacha de un gesto" [breasts where sea breezes nurse / fainting lemon tree / moribund lemon tree. / Lemon Tree. You / saw my love / broken by the ax of a gesture] (303).

24. Lorca, *Epistolario*, 2: 78–79.

25. Ewing, *The Body*, 22.

26. In a 1925 drawing by Lorca, Dalí is depicted with one of his hands in a fishbowl with a fish on each finger and a larger fish on his chest.

27. For the original version and English translation of "Three Portraits with Shading," see Lorca, *Collected Poems,* ed. Maurer, 452–58.

28. Federico García Lorca, "Sketch de la nueva pintura," *Conferencias,* 2: 45.

29. For Mario Hernández the theme of the bath of love comes from the Sephardic *cancionero* and from popular songs; see "La muchacha dorada por la luna," in *Trece de nieve,* nos. 1–2 (December 1976), 212.

30. Lorca, *Obras completas,* 21; *Collected Poems,* 591.

31. In that sense, Lolita is a reminder of *Estrellita de oro* [Little Golden Star], an ideal female figure in his youthful *Book of Poems,* embodiment of childhood dreams and faith who becomes *Esfinge de nieve* [Sphinx of Snow], synonymous with the moon, when the poet loses faith and love with the passage of time. For a detailed study of these early images, see the first part of Candelas Newton's *Una escritura en trance: Libro de poemas y Diván del Tamarit* (Philadelphia: Benjamins, 1992).

32. Ewing, *The Body,* 32, 34.

33. Ibid., 36.

34. Ibid., 37.

35. Hutcheon, *Theory of Parody,* 6.

36. Barbara G. Walker, *The Woman's Encyclopedia of Myths and Secrets* (San Francisco: Harper & Row, 1983), 117.

37. Lorca, *Epistolario,* 2: 79. In his essay on Saint Sebastian, a figure both Lorca and Dalí greatly admired, reprinted in Rafael Santos Torroella, *Los putrefactos de Dalí y Lorca: Historia y antología de un libro que no pudo ser* (Madrid: CCSC, 1995), the painter wrote: "On the top part of the heliometer was the multiplying glass of Saint Sebastian. This glass was concave, convex and flat at the same time. Recorded on the platinum cover of its clean and exact crystals, the following inscription could be read: 'Invitation to Astronomy' and below, with letters which imitated the relief: 'Saint Objectivity.' On one numbered crystal rod, this was still legible: 'Measurements of the apparent distances between pure esthetic values.' "

Luis Buñuel as Intertext in the Narrative Prose of Juan Goytisolo

Virginia Higginbotham
University of Texas, Austin

FOR Luis Buñuel, surrealism, among the most successful of the prewar "ismos" of modernist art and literature, provided a strong signifying system with which he created films whose techniques are still considered landmark experiments. Surrealism also inspired other members of Spain's Generation of 1927, including some of the country's finest poets, among them—García Lorca, Rafael Alberti, Luis Cernuda, and Vicente Aleixandre—who experimented with chaotic imagery and syntax, as well as *postista* poets of the Spanish postwar era such as Edmundo de Ory and Miguel Labordeta. Seventy-five years after the publication of the first *Manifeste du surréalisme* (1924) in Paris, surrealism as a literary language retains its force not only in Spanish poetry but also in the narrative fiction of Spanish prose writers such as Juan Goytisolo. In a 1972 interview reprinted in *Disidencias*, Goytisolo expressed his admiration for Buñuel's films, specifically *Discreet Charm of the Bourgeoisie, The Milky Way,* and *Simon of the Desert.*[1] My brief review of the relationship that resonates between Buñuel's films and some of Goytisolo's mature prose texts is based upon Roland Barthes notion of intertexuality as a "plurality of other texts"[2] and Julia Kristeva's discussion of the intersection of texts where at least one other text can be read.[3] My study intends to establish how surrealism as a system of signification and film as a narrative device, both represented in the cinema of Luis Buñuel, form a juncture of two significant intertexts in the mature narrative prose of Juan Goytisolo.

Early in his career Goytisolo acknowledges how film has changed the way we perceive and narrate stories: "En efecto, en 1956 no percibimos las cosas del mismo modo que hace cuarenta años: hemos adquirido la costumbre de VER CONTAR historias, en lugar de oirlas narrar a la manera de las novelas del siglo XIX" [In effect, in 1956 we don't perceive things in the same way as we did forty years ago: we have acquired the habit of watching stories being told instead of hearing them narrated as nineteenth-century novels were].[4] Awareness of the visual nature of film and its language is interwoven on several levels in Goytisolo's texts: Filmic characters such as King Kong emerge as narratorial alter egos—"el tú interpelado se confunde sucesivamente con los protagonistas de *King Kong, Locura de amor,* y *Lawrence de Arabia*" [the interpolated "tú" is confused successively with the protagonists of *King Kong, Love Madness,* and *Lawrence of Arabia*) (*D,* 325)—and references to film directors (Hitchcock, Bergman, Antonioni, in addition to Buñuel) are as frequent in his narratives as references to literary works. Use of English terms such as "on-screen," "the reel stopped turning," and references to voice off, indicates the scriptwriter's familiarity with film production and is integrated into the narration of *La reivindicación del conde don Julián* (The revindication of Count Julian, 1970). In *Señas de identidad* (1966), the narrator's very identity—that of a filmmaker who has returned to Spain to film a documentary on the emigration of Spanish workers—seems to follow a Buñuelian model. For, like Buñuel, Goytisolo is never nonideological: in the same early essay, Goytisolo establishes very clearly the importance of the writer's point of view: "Narrar es tomar una posición—un ángulo de enfoque . . . sobre lo que se narra" [To narrate is to take a position—a point of view . . . about what is narrated] (*P,* 22). Goytisolo's early fictional travelogues leave no doubt that this postwar writer shares the point of view and the same corrosive, satirical critique of society and its hypocrisies that Buñuel inherited from surrealism.

An early example of Buñuel's ideological cinema is *Las Hurdes: Tierra sin pan* (Las Hurdes: Land without bread, 1932), a combination of documentary and social exposé punctuated by the kind of jolting juxtapositions with which surrealists shocked and amused their audiences. Buñuel and his film crew of three discovered that in the rocky, poverty-ridden region of Las Hurdes, only four hours from Madrid by car, there existed inhabitants suffering from various diseases known to be treatable and were so poor that bread was unknown. Buñuel's documentary records the lives of the Hurdanos in their celebrations, their struggle to survive, and their efforts to educate their children. In the schoolroom we see barefooted children who do not even own shoes write on a blackboard the motto of a consumer society that is entirely irrelevant to their own: "Respect the property of others."[5] The presence in Las Hurdes of disease is emphasized by repeated images of suffering: a woman, only thirty-two years old with goiter, a child with infected teeth and gums, an old man shivering with fever on a hillside, a man bitten by a snake.

These images underscore the indifference of affluent urban societies to the suffering of the less fortunate by underlining the fact that diseases such as malaria, contained elsewhere, remain in pockets of neglect such as Las Hurdes.

Understood through *Las Hurdes* as intertext, Goytisolo's travel documentaries read like sequels to the 1932 Buñuel documentary. Las Hurdes is substituted by the Andalusian province of Almeria, another of Spain's unproductive and poverty-stricken provinces. In *Los campos de Níjar* (The Fields of Nijar, 1959) the narrator encounters workers isolated from progress by centuries of poverty and still exploited by absentee land owners who profit from rents paid by penniless occupants. Like Buñuel in *Las Hurdes,* the narrator of *Los campos de Níjar,* as Randolph Pope explains, is moved by what he sees and becomes an "advocate, a medium for an ignored netherworld."[6]

Buñuel's mixture of social outrage controlled within parameters of documentary film narrative is articulated by Goytisolo with fictional characters in *La Chanca* (1962), an urban suburb of Almeria. Like the narrator of *Las Hurdes,* the narrative voice of *La Chanca* comes from the outside, this time from Paris, to visit a region perceived as a relatively exotic land. In the area of La Chanca, as in Las Hurdes, the inhabitants are reduced to a sordid existence, exploit each other for profit, and live surrounded by corruption and sickness. For Pope the presence of disease implies a "comparison of the motherland with a body and of moral abasement with physical sickness" (*UGJ,* 71), an echo of Buñuel's portrayal of Las Hurdes as a land whose enormous potential is prostituted by outsiders, leaving its inhabitants ill, ignorant, and dependent upon their exploiters for their very livelihood. Goytisolo was planning another such fictional documentary, *La isla* (1960), this time in the form of a film script. An exposé of the decadence of wealthy Spanish residents in Torremolinos, its filming was prohibited by the Franco government so it was published as a novel that denounces the materialism and uselessness of the haute bourgeoisie.

Buñuel's *Las Hurdes* and Goytisolo's fictional documentaries represent their creators' early preoccupation with the *abject,* a term Julia Kristeva defines as the "violence of mourning for an 'object' which has always already been lost." "The abject," Kristeva continues, "shatters the wall of repression and its judgments . . . Abjection is a resurrection that has gone through death (of the ego). It is an alchemy that transforms the death drive into a start of life, of new significance."[7] Buñuel's documentary and Goytisolo's fictional documentaries are abject in the sense that their violent imagery attempts to recall whole populations who have been forgotten, exploited, or repressed by the more fortunate.

Both Buñuel and Goytisolo find relief from the abject, as well as a reminder of it, in dark humor, which André Breton celebrated in his *Anthologie de l'humour noir* (1940). Surrealist humor is subversive, the kind described by Kristeva in her discussion of Menippean discourse as a "desecration of the sacred" (*KR*, 83). Both Buñuel and Goytisolo are masters of counterdiscourse in general, and of humor in particular. As a counterdiscourse designed to undermine the dominant discourse and thus liberate themselves (and us) from it, Buñuel and Goytisolo's humor is ironic, frequently scatological, almost always directed against figures of authority and the conformist rituals practiced by members of the social class they both dropped out of, the bourgeoisie. The violence in *L'Age d'or* with which Modot slaps the Marquise for spilling drops of a cocktail on his trousers is so excessive that it surprises and evokes laughter in the same way as Goytisolo's grotesque treatment of Mrs. Potiphar, the grande dame tourist whose false dignity is belied by her seersucker suit and sunglasses.[8]

Humor, Michael Rifaterre explains, is a special case of poetic language that functions as a variety of intertextuality which posits structural polarities.[9] Such juxtaposition of opposite terms is one of surrealism's fundamental poetic devices and is frequent in Buñuel's films, for example, the image of a laughing Jesus in *Nazarín*. This image shocked religious viewers because of its implied juxtaposition of the suffering face of Jesus of traditional Christian iconography with an image that contradicts centuries of Christian visual iconography in its assertion that Jesus may have laughed, a suggestion entirely absent from Christian representation. The narrator of *Don Julián* enjoys a similar parody of the austere icons of Spain—Seneca, the writers of the Generation of 1898, even that of Don Quijote himself, "ranciosa cepa senequista, Garbanzote de la Mancha" [Seneca's rancid stock, garbanzo bean of La Mancha] (*DJ*, 151), and the parody of the perfect Spanish gentleman, Don Alvaro Peranzules whose solemn visage ("su rostro, severo y enjuto") [his face, severe and skinny] (*DJ*, 159) personifies the gravity of his generation (of 1898) as he sits reading Calderón and Lope de Vega.

The Christian preoccupation with spiritual purity has often led to the church's refusal to make allowances for human corporeality. The lengths to which the church has often gone to avoid acknowledging bodily functions form the basis for jokes by surrealists and diatribes by Goytisolo. Buñuel's famous portrayal in *Phantom of Liberty* (1974) of guests socializing while seated on commodes, yet who rush out surreptitiously to eat in private, ridicules Christian notions of purity by reversing the manner in which we eat and defecate. In the same way Goytisolo recalls Alvarito's consternation: "la idea de la profanación sacrílega le abrumaba de tal modo que . . . la materia fecal no alcanzaba a abrirse paso" [the idea of sacrilegious profanation overwhelmed him to such an extent that his fecal matter couldn't manage to make its way out].[10] Goytisolo extends his satire of the notion of purity even futher, for Alvarito personifies "un país (el suyo) . . . estreñido" [a country (his) that is constipated] (*JST,* 226). Humor based on food and excrement, Kristeva points out, is part of the structure of the carnivalesque, a complex, antitheological, popular discourse in which laughter "is no more comic than tragic; it is both at once" (*KR,* 50).

Buñuel's films focus upon a parade of obsessed and criminal characters, from the furious Modot in *L'Age d'or* whose frustration knows no bounds (he kicks a dog, slaps the Marquise, throws a plow out the window) to the pathological Francisco of *El,* who suspects his innocent wife of infidelity, to Archibaldo de la Cruz, who has no clue until the end of the film as to why he is driven to kill women. Also memorable are the murderous priest and the drug-peddling ambassador of Miranda in *The Discreet Charm of the Bourgeoisie.* Among the funniest of these corrupt and pathological figures is Simon of the Desert, Buñuel's recreation of the historical third-century Syrian priest, Simon Stylite, who displayed his ascetic devotion by living atop a tall column in the desert. Buñuel plays with the numerous contradictions inherent in asceticism: Simon, residing on his column, abstains from food, trying unsuccessfully to resist temptation. He reaches out in ecstasy as a vision of Christ carrying a lamb appears, only to see the Lord turn into an attractive blond woman. According to early Christian belief that sin originates with women, Satan is disguised in female form. Real women, such as his mother, Simon receives with disdain, for they detract from his obsession with his own saintliness. It is Buñuel's understanding of the similarities between desire and obsession that makes his Simon laughable, and that Goytisolo celebrates when his wandering narrator assumes various roles including those of Cavafy and Lawrence of Arabia and realizes: "Has . . . emulado la hazaña del Estilita en lo alto de la buñuelesca columna" [You have . . . emulated the heroic feat of the Stylite atop the Buñuelesque column] (*JST,* 174). This direct allusion to *Simon of the Desert* confirms an intertextual relationship with Buñuel and enlarges Goytisolo's text by the addition of a range of associations with Buñuel's film.

Caricature merges easily into the grotesque, a mode deeply ingrained in Spanish art (Goya) and literature (Quevedo) long before the appearance of surrealism, and which continues in twentieth-century Spanish film (Buñuel, Saura, Almodóvar) and literature (Valle Inclán, Martín-Santos, Goytisolo) not so much as intertext as the articulation of a national discursive mode. Components of the grotesque include the horrific, for which surrealism has a special proclivity, as Rosa Mª Martín Casamitjana explains: "El Surrealismo, al indagar en las zonas ocultas de la mente y desenterrar las pesadillas, erige mundos habitados por criaturas horrendas, semejantes a las soñadas por Bruegel y el Bosco" [Surrealism, on dredging the hidden regions of the mind and unearthing nightmares, raises worlds inhabited by horrendous creatures, similar to those dreamed up by Brueghel and Bosch].[11]

Buñuel's contribution to the imagery of the Spanish grotesque is extensive, from his shot of skeletons still dressed in their bishops' miters bleaching on the rocks of Mallorca in *L'Age d'or* to the beggar dressed in a bridal veil mocking the vows of marriage to Christ that Viridiana plans to take when she returns to her convent. Goytisolo's grotesqueries are too extensive to detail here but undoubtedly include the memorable specter of death/aids in *Las virtudes del pájaro solitario* (Virtues of the solitary bird, 1988). The personification of disease, this figure, female, as is Simon's vision of Satan, is both Goyesque—small dolls hanging from her hands are reminiscent of Goya's *The Witches Sabbath*—as well as Daliesque with her elongated "filiform" extremities recalling the spidery legs of the animals in Dalí's *Temptation of Saint Anthony.* The figure pictured on the jacket of this book, however, is by the Belgian painter Felicien Rops.

Buñuel exploits the special capacity of the grotesque to shock and repel with repeated images of insects in his first three films. Dalí and Buñuel's shared fascination with the symbolic value of insects is evident in the now-famous image of ants swarming in the palm of the young man itching with erotic desire in *Un Chien andalou.* One of the few close shots of *L'Age d'or* is the repelling scene of the struggle between a rat and a scorpion in the opening frames that prefigures the violence of the film. A more detached textbook illustration of the anopheles mosquito in *Las Hurdes: Tierra sin pan* is a welcome relief from Buñuel's close-ups of insect life. Insects as emblems of fear appear to be icons of a generational language, as they are prominent in poems by Lorca such as "Luna y panorama de los insectos" in *Poeta en Nueva York,* as well as in Dámaso Alonso's postwar "Los insectos" in *Hijos de la ira* (Children of ire).

This repeated sign of terror, apparently a favorite of Lorca, Dalí, and Buñuel, provides a similar moment of horror in *Don Julián* when the young Julian goes to the library and smashes a collection of insects between the pages of works by classical Spanish authors (*DJ*, 37). Sheer terror strikes Julian in a biology class lab session when treated by the professor to the "modesto drama" of a grasshopper devoured by a scorpion in a glass jar (*DJ*, 93). When the pompous academic, Don Alvaro Peranzules, rises to address the Royal Academy on the overworked topic of "Consideraciones en torno al concepto del honor en el teatro español del siglo XVII" (Considerations about the concept of honor in the seventeenth-century Spanish theater), the smashed insects revive, crawl from the "mohosas páginas" [moldy pages] of the volume from which Don Alvaro reads to the Academy, and begin to fly about: "los insectos zumban, vuelan, copulan, se multiplican, buscan alimento en los libros magistrales de los estantes" [the insects buzz about, copulate, multiply, feed on the masterworks on the shelves] (*DJ*, 180). The visual power of this scene and its filmic potential represents another link in the close relationship between Goytisolo's text and Buñuel's cinema. The defilement of the classics spills over onto the august fellow of the Royal Academy who is attacked by the insects emerging from the volumes, shrinks up and dies.

If Buñuel's surrealist films raise grotesque humor to new heights, they also articulate a topic of concern to surrealists, but one that is difficult to visualize, that is, the theme of the "other," or the double. In *Interpretation of Dreams* Freud wrote that dream formation is possible because sleep reduces the censoring self.[12] In *Un Chien andalou* Buñuel visualizes the censoring self, a manifestation of the "other," with great clarity when the young protagonist confronts an angry young man who looks just like himself. The censoring "other" chastises the protagonist and turns him to the wall as a parent punishes a child. Finally the protagonist rebels against his parental "other" and "kills"/shoots him when a pair of books on a desk turn into a pair of revolvers. This division of one personality into opposing components, evident in Buñuel's first film, is also a theme of his last, *That Obscure Object of Desire.* Here, the two opposing facets of the personality of a young woman, Conchita, are so distinct as to be played by different actresses.

The theme of the double so pervades Goytisolo's later prose that most of his characters are multifaceted, have several names and personas: in *Juan sin tierra* Doktor Vosk is also Colonel Vosk as well as Padre Vosk; Mrs. Potiphar is also "Hija de la Revolución Americana," dies and rises again(*DJ*, 69). The double character is also prominent in *Señas de identidad*, a novel whose central issue is the protagonist's search for identity. In it, the protagonist, Alvaro Mendiola, tries to recall events surrounding his father's assassination by Franco's troops. He researches news accounts of the events, but his memory, saturated with photographs and separated by time and distance from the event, becomes murky. He experiences a strange "desdoblamiento" in which he sees the event both "como actor y como testigo, espectador, cómplice y protagonista a la vez del remoto y obsesionante drama" [as actor and witness, spectator, accomplice, and protagonist of this distant and obsessive drama all at the same time].[13] If Alvaro can no longer distinguish between truth and fiction, his very identity seems to fuse with that of his best friend, Antonio, recently released from prison. Antonio shares Alavaro's thoughts and views so completely that, according to Christian Meerts, Antonio "n'est qu'une variante . . . un reflet d'Alvaro" [is only a variant . . . a reflection of Alvaro].[14] Thus Goytisolo uses the figure of the double to represent a composite of the revolutionary: Alvaro, the intellectual residing in Paris, and his alter ego, Antonio, the activist in Spain. These two facets of the protagonist's character are in harmony rather than in conflict, as in *Un Chien andalou.*

Surrealists found the theme of the double to be multidimensional, capable of representing inner conflict as well as the search for the "other" representing love, desire, and passion. As Raquel Medina points out, "el amor se define como la búsqueda del "otro" [love is defined as the search for the "other"].[15] The fusion of the "I" with a beloved is the flip side of the theme of the censoring "other." The "other" thus can introduce love and erotic desire, themes central, not just to surrealists, but which surrealists intensified with the term "l'amour fou" [mad passion]. Surrealists attached supreme importance to *l'amour fou* as the only source of personal energy intense enough to change human behavior. In his essay, *L'Amour fou* (1937), Breton emphasizes the surrealist notion that love should be free and spontaneous when he declares: "Il n'y a jamais eu de fruit défendu" [There has never been a forbidden fruit].[16] Buñuel pursues this notion when he explores the demonic character of erotic passion in *Cumbres borrascosas* (Stormy heights, 1953). This film is a Mexicanized version of Emily Brönte's *Wuthering Heights* that pays tribute to the theme of passion beyond the reach of logic celebrated in the Gothic novel. The theme of *l'amour fou* is one of the most consistent of Buñuel's films and explains his repeated return to the hopeless fixation of older men with younger women in *Viridiana, Tristana,* and *That Obscure Object of Desire.*

In Goytisolo's texts *l'amour fou,* passion beyond reason, becomes one of the writer's marks of identity. In *Juan sin tierra,* the narrator articulates as his own a motto echoing the surrealist recognition of the power of human instincts: "acamparemos allá adonde nos lleve el instinto" [we will follow where our instinct leads] (*JST,* 89). Goytisolo is particularly intrigued by associations of passion with acts of power, vengeance, deceit, and the demythologizing of cultural and religious codes. Goytisolo's sexuality is associated with his identity as a writer and reaches the level of textuality in *Juan sin tierra* in which the writer equates writing with onanistic pleaure, "placer solitario de la escritura!" [the solitary pleasure of writing!]. He also enjoys more predatory pursuits, imagining himself a bird of prey swooping into the "urban jungle" where he will be drawn to his target "con fuerza magnética como palabras unidas por el poeta en versátil conjunción fugaz" [with magnetic force like words joined together in flexible, fleeting union] (*JST,* 298). Use of the term "fuerza magnética" inevitably links Goytisolo's conception of sexual attraction to Breton's description in *L'Amour fou* of desire as an uncontrolled magnetism, "La sympathie qui existe entre deux, entre plusieurs êtres" [the sympathy existing between two or several beings] (*AF,* 50). The fact that Goytisolo's narrator practices homosexual passion gives a tone to the topic unintended by Breton, but such a transposition does not appear to alter the conception of "mad passion" or the vocabulary Goytisolo uses to describe it.

Breton's first surrealist manifesto of 1924 was, among other things, an opening salvo against logic and its literary manifestation, the realist novel, its minute descriptions of the materialist world, and its propensity for psychological analysis as opposed to revelation. Rather than mirror the outer world, surrealists hoped to reveal suppressed desires and uncensored thoughts of the inner world, to add the insights gained from irrationality and dream to those of the rational, ordered mind and thus to escape what Breton called "le règne de la logique" [the rule of logic].[17] Buñuel in his first two films destroys the structures of narrative logic— chronological sequence, characterization, psychological analysis. The result destabilizes meaning and challenges confused viewers to sort out the film's clues. The attack on the grammaticality, or logical structure, of language is among the most significant of surrealism's intertextual contributions to Goytisolo's writing, for it provides a model for his own project of renewal of literary/artistic language. As an exile in Paris who began his writing career as a realist, Goytisolo complained to Claude Couffon in an interview in 1970 that most Spanish writers committed to critique the ruling class did so in the language of that class: "Je n'ai absolument pas renoncé à l'engagement. Je l'ai simplement transféré sur un autre plan" [I absolutely have not renounced commitment. I have simply transferred it to another level].[18]

The language of dream became for surrealists a model of "automatic" or uncensored writing that seems chaotic but which offers intriguing if mysterious series of words and images. Buñuel's first two films as well as some later ones are examples of dream language in which apparently unrelated images are juxtaposed without explanation, conventions of logical syntax forgotten, and images metamorphosing into others in anarchical profusion. The illogic of *Un Chien andalou* and *L'Age d'or* is never as rampant in Buñuel's later films, when he realized he wanted to make films as a profession and must communicate more effectively with a wide public, although he never ceased to make use of the devices of dream language.

Goytisolo came late to the language of chaos, but in *Señas de identidad* realism begins to loosen its grip on his writing. Just as Buñuel, with the still-shocking image of a razor slicing an eyeball, suggests that we see without preconceptions, Goytisolo begins to discard the trappings of logical discourse by writing pages of prose unencumbered by punctuation or construction of realistic fictional characters. As Abigail Lee Six points out, Goytisolo in his post-*Señas* fiction sets about the destruction of language with a vengeance: "We can take nothing for granted: not the meanings of words and set expressions, not even conventions of punctuation and lay-out."[19] The narrator of *Juan sin tierra* suggests that this demolition project is part of the dynamics of redefining his cultural identity: "no basta con echar por la borda rostro, nombre, familia, costumbres, tierra: la ascesis debe continuar: cada palabra de su idioma te tiende igualmente una trampa: en adelante aprenderás a pensar contra tu propia lengua" [it is not enough to throw overboard face, name, family, customs, homeland: the process must continue: each word of your language entraps you: from now on you will learn to think against your own language] (*JST,* 83).

Buñuel's creation of narrative confusion in *The Discreet Charm of the Bourgeoisie* (1972) especially intrigued Goytisolo, who admits in an interview that he learned an important strategy from this film:

> *El discreto encanto de la burguesía* me sedujo por muchas razones; pero la que me impresionó quizá más fue su utilización habilísima de esa técnica del engaste, tan común en la narrativa precervantina, que suspende el desarrollo de la acción del modo más arbitrario e inmoviliza a los personajes como estatuas de sal, conviertiéndoles en simples receptores de la palabra del "hombre-relato." El episodio de los militares que irrumpen en la cena y aplazan su retorno a las maniobras a fin de oír referir el sueño de uno de sus camaradas reúne todas las características de lo que la poética denomina hoy "relato inmotivado": inverosimilitud, personajes "intransitivos," carentes de espesor sicológico.

[*The Discreet Charm of the Bourgeoisie* attracted me for many reasons; but the one which impressed me most perhaps was his very skillful use of the technique of setting up situations, so common in precervantine narrative, which suspends the development of the action in the most arbitrary way and immobilizes the characters like pillars of salt, turning them into receivers of the word of the storyteller. The episode of the military troop that interrupts a dinner and delays its return to maneuvers until hearing one of its comrades tell his dream joins together all the characteristics of what poetics today calls the "unmotivated story": contrivance, lacking in psychological substance.] (*D*, 324)

Goytisolo's thoughts on narrative, quoted here in full in order to clarify the intertextual exchange between prose and film narrative, are inspired by Buñuel's *Discreet Charm of the Bourgeoisie,* the random or chance appearance of its narrative, and the director's ability to create spontaneity by means of surprise and intrigue. In *Las virtudes del pájaro solitario* Goytisolo applies these devices of surrealist film narrative to what appears to be unstructured prose. Both the identity of characters: "él es yo? yo soy ella?" [he's me? am I she?] as well as their gender shift unreliably in this book.[20] As in *The Discreet Charm of the Bourgoisie,* the narrative of *Pájaro solitario* is framed by the presence of a set of characters who serve as a collective sign connecting a digression to the primary narrative. Buñuel's characters have names, but their significance is as members of a social class rather than as individuals. Goytisolo's characters are unnamed residents of a hotel/sanatorium who serve the same narrative function: that of a sign restarting the narrative. Buñuel's habit of framing confusing or bizarre scenes as if they are segments of a dream is also employed by Goytisolo in *Pájaro solitario.* He introduces a sequence of action as evolving from within multiple dreams, "como en un sueño dentro de un sueño dentro de un sueño, pero enteramente despierto percibía . . . la sorda marea de voces" [as in a dream within a dream within a dream, but entirely awake, I perceived . . . the sound of murmuring voices] (*PS*, 77).

Buñuel often proclaimed that he was an atheist, yet he confounded critics and catholics alike by treating religious topics in his films. *Nazarín* (1958) portrays a dissident priest, and *The Milky Way* (1969) recreates the historical debates that have arisen about six controversial points of catholic dogma. This remarkable film also intrigued Goytisolo with its narrative devices: "Es muy interesante ver cómo Buñuel incorpora a su obra una gran variedad de técnicas narrativas (piensa, por ejemplo, en el ritmo acelerado del duelo ideológico, a punta de espada, entre el jesuita y el jansenista en *La vía lác-*

tea") [It's very interesting to see how Buñuel incorporates into his work a great variety of narrative techniques (think, for example, of the accelerated pace of the ideological duel, at sword point, between the Jesuit and the Jansenist in *The Milky Way*)] (*D*, 324). The scene to which Goytisolo refers is one in which there is a kind of counterpoint between the eighteenth-century combatants' verbal arguments and the thrust and parry of their swords as they engage in a duel. Thus Buñuel makes clear that wars of dogma are fought on two levels and that ideological belligerance can be as fatal as the point of a sword.

Just as Buñuel admires Nazarín as a rogue priest who rejects church dogma and dedicates his life to service to others, Goytisolo also is inspired by a religious rebel. Having the narrator of *Juan sin tierra* squash flies and scorpions in the pages of the Spanish classics as a sign of their depletion, Goytisolo now pays tribute to one of Spain's most revered sixteenth-century writers, the mystical poet San Juan de la Cruz, in *Pájaro solitario.* Like the atheist Buñuel who made films about a priest and catholic dogma, Goytisolo identifies with San Juan, a persecuted writer imprisoned by his own religious sect for his efforts at reform. To remove incriminating evidence, San Juan is said to have destroyed some of his important texts by eating them. Among his lost texts was one entitled *Las virtudes del pájaro solitario,* which Goytisolo now reconstructs on the basis of research showing links between Christian and Islamic religious icons, one of which is a solitary bird.

Dividing Spanish fiction and film into two phases, the mimetic, or representational, and the intradialogic, Gonzálo Navajas writes that "Spanish fiction and film of the intradialogic phase are not characterized by one-to-one correspondences between aesthetic objects and their well-defined referents but by multiple sources in complex webs of signifying relations."[21] To no writer is this comment more applicable than Juan Goytisolo, whose texts are indeed "complex webs" or "a mosaic of texts" to use Kristeva's metaphor. Goytisolo acknowledges a considerable list of literary "ancestors and alter egos" which Pope numbers to be about twenty-nine (*UJC*, 147). I suggest that, more than an alter ego, Buñuel's surrealist films are direct intertexts of Goytisolo's fiction, a possibility critics have yet to investigate. For, besides Goytisolo's specific references to his films in an interview, and allusions to the "Buñuelesque column" in *Juan sin tierra,* the traces of Buñuel's humor, imagery, and filmic narrative structures are recognizable in Goytisolo's ambiguous, indeterminate, metamorphosing prose. Most importantly, Goytisolo learned from Buñuel's surrealist films, as Kristeva says of some oppositional discourse, "to make the conscious/unconscious distinction irrelevant" (*PH,* 7). In doing so Goytisolo's prose achieves the surrealist goal of uniting the rational and the irrational. Another surrealist, Antonin Artaud, whom Goytisolo cites, hoped that surrealism could bring about the fusion of poetry and anarchy.[22] In his late prose, Goytisolo fulfills this hope.

Notes

All translations are my own unless otherwise indicated.

1. Juan Goytisolo, *Disidencias* (Barcelona: Seix Barral, 1978), 324–25. Hereafter *D,* cited in the text.

2. Roland Barthes, *S/Z: An Essay,* trans. Richard Miller (New York: Farrar, Straus & Giroux, 1974), 10.

3. *The Kristeva Reader,* ed. Toril Moi (New York: Columbia University Press, 1986), 37. Hereafter *KR,* cited in the text.

4. Juan Goytisolo, *Problemas de la novela* (Barcelona: Seix Barral, 1959), 21. Hereafter *P,* cited in the text.

5. For detailed discussions of Buñuel's films, see Virginia Higginbotham, *Luis Buñuel* (Boston: Hall, 1979); Gwynne Edwards, *The Discreet Art of Luis Buñuel* (London: Boyars, 1982); Agustín Sánchez Vidal, *Luis Buñuel: Obra cinematográfica* (Madrid: Ediciones J.C., 1984).

6. Randolph Pope, *Understanding Juan Goytisolo* (Columbia: University of South Carolina Press, 1995), 68. Hereafter *UJG,* cited in the text.

7. Julia Kristeva, *The Powers of Horror: An Essay on Abjection,* trans. Leon S. Roudiez (New York: Columbia University Press, 1982), 15. Hereafter *PH,* cited in the text.

8. Juan Goytisolo, *La revindicación del conde don Julián* (Mexico: Mortiz, 1966), 66–67. Hereafter *DJ,* cited in the text.

9. Michael Rifaterre, "Poetic Functions of Intertextual Humor," *Romanic Review* 65 (1974): 293.

10. Juan Goytisolo, *Juan sin tierra* (Barcelona: Seix Barral, 1975), 116. Hereafter *JST,* cited in the text.

11. Rosa Mª Martín Casamitjiana, *El humor en la poesía española de vanguardia* (Madrid: Gredos, 1996), 124.

12. Sigmund Freud, *The Interpretation of Dreams,* trans. A. A. Brill (New York: Random House, 1950), 383.

13. Juan Goytisolo, *Señas de identidad* (Mexico: Mortiz, 1966), 116. Hereafter *S,* cited in the text.

14. Christian Meerts, *Technique et vision dans "Señas de identidad" de J. Goytisolo* (Frankfurt: Klosterman, 1972), 18.

15. Raquel Medina, *El surrealismo en la poesía española de posguerra (1939–1950)* (Madrid: Visor Libros, 1997), 113.

16. André Breton, *L'Amour fou* (Paris: Gallimard, 1937), 136. Hereafter *AF,* cited in the text.

17. André Breton, *Manifeste du surréalisme* (Paris: Gallimard, 1973), 18.

18. In Meerts, *Technique et vision,* 56 n. 267.

19. Abigail Lee Six, *Juan Goytisolo: The Case for Chaos* (New Haven, Conn.: Yale University Press, 1990), 36.

20. Juan Goytisolo, *Las virtudes del pájaro soltario* (Barcelona: Seix Barral, 1988), 58. Hereafter *PS*, cited in the text.

21. Gonzálo Navajas, "Intertextuality and History," in *Intertextual Pursuits: Literary Mediations in Modern Spanish Narrative,* ed. Jeanne P. Brownlow and John Kronik (Lewisburg, Penna.: Bucknell University Press, 1998), 151.

22. Juan Goytisolo, *El furgón de la cola* (Barcelona: Seix Barral, 1967), 90.

Romanticism, Transcendence, and Modernity in Lorca's *Libro de poemas*, or the Adventures of a Snail

Michael Iarocci
University of California, Berkeley

NOSTALGIA for the innocence of childhood, the disenchantment of the world, the confused awakenings of adolescent desire; such are the thematics most often invoked in critical accounts of Federico García Lorca's first book of poems, the 1921 *Libro de poemas* (Book of Poems). Indeed, Lorca himself established this interpretive frame explicitly in the "Palabras de justificación" (Words of justification) that introduce the collection: "ofrezco en este libro, todo ardor juvenil y tortura, y ambición sin medida, la imagen exacta de mis días de adolescencia y juventud, esos días que enlazan el instante de hoy con mi misma infancia reciente" [I offer in this book, which is entirely youthful ardor, torture and measureless ambition, the exact image of my days of adolescence and youth, those days that bind the present instant with my recent infancy itself].[1] Not surprisingly, it has been through this prism of recollected youth that some of the more salient features of the work's poetic landscape—that is, the familiar, enchanted sense of place, the magical animal world, the lyric subject's intercourse with his surroundings, etc.—have been understood.[2]

But if this neoromantic topos provides a relatively comfortable hermeneutic cipher for the *Libro de poemas*, it also opens up a series of questions that have received less critical attention. They are those hallmark questions that have persistently engaged critics of an older romantic rhetoric, questions pertaining to the relationship between the lyric subject and time.[3] In the case of Lorca's *Libro de poemas*, the question of time would seem to be fundamental, inasmuch as it bears not only on the individual poems, but also more generally, on the structure of the book itself. For if framing the collection as a poetic register of lived experience invites inquiry into the specificity of the "past moments" it conveys, it simultaneously calls attention to the sequence, syntax, and trajectory of those moments. It suggests that in addition to the intricate nature of the lyric "present" of each poem, the *Libro de poemas* might also be read profitably (within the long legacy usually traced to the Petrarchan *canzionere*) as a vaguely temporal, pseudo-narrative progression, as a succession of psychobiographical states. The "Palabras de justificación," coupled with the loosely chronological arrangement of the poems would thus initially convey to readers of the book the sense of movement through biographical time.

This suggestion of temporal movement across poetic texts seems to find a rather explicit thematic analogue in the travel motif that pervades the work. "Journey" and its constellation of related terms—paths, streets, walks, views, escapes, etc.—constitute an important semantic cluster of the *Libro de poemas.* It is a facet of the work that, again, Lorca himself seemed to anticipate in his introductory words when he observed that, among the book's virtues was its ability to permanently remind him of his "infancia apasionada correteando desnuda por las praderas" [impassioned childhood running naked through the meadows]. Indeed, a brief glance at the poems of the *Libro de poemas* reveals the persistence with which the image of the lyric subject as a traveler-wanderer manifests itself in individual compositions. The following are representative samples, each from a different poem: "mi senda se pierde en el alma de la niebla" [my path is lost in the soul of the mist] (109); "voy camino de la tarde entre flores de la huerta" [I walk toward the afternoon among the flowers of the orchard] (111); "yo / voy llorando por la calle" [I go crying along the street] (112); "La sombra de mi alma / huye por un ocaso de alfabetos" [My soul's shadow flees through a sunset of alphabets] (123); "¡Voy en busca de magos / y de princesas!" [I go in search of wizards and princesses!] (179); "Yo me salgo desnudo a la calle / maduro de versos / perdidos" [I go out into the street naked, matured by lost verses] (182). From the problematic search for direction in the first poem, "Veleta" (Weather vane) to the lustful wanderings represented in the last, "El macho cabrío" (The he-goat), the leitmotif of journey is evident. Moreover, to the extent that travel is among the more traditional topoi for the representation of the human "journey" through time, it is not difficult to appreciate the metapoetic dimension of this motif. Images of travel thus figure the temporal progression loosely encoded in the passage from one poem to the next.

Within this context, "Los encuentros de un caracol aventurero" (The encounters of an adventurous snail), the book's second poem, which narrates a snail's odyssey out into the world, offers a unique point of entry into the *Libro de poemas*. As one of the opening poems of the book, and as the only composition structured entirely as a travel narrative, "Los encuentros" occupies a privileged position within the collection. At the same time, the metapoetic dimension of the travel motif suggests that this poem's particular problems, questions, and dynamics might provide insight into the work as a whole. In the following pages I will consequently focus on "Los encuentros de un caracol aventurero" 1) in order to understand more fully the voyage the poem charts and 2) in order to explore its relationship to the broader journey that is the *Libro de poemas* itself.

The poem opens with a modernized yet readily identifiable *locus amoenus*:

> Hay dulzura infantil
> en la mañana quieta.
> Los árboles extienden
> sus brazos a la tierra.
> Un vaho tembloroso
> cubre las sementeras,
> y las arañas tienden
> sus caminos de seda
> —rayas al cristal limpio
> del aire—. En la alameda
> un manantial recita
> su canto a las hierbas.

[There is a childish sweetness in the still morning. The trees spread their arms to the earth. A trembling vapor covers the fields, and the spiders hang their silk paths—lines across the air's clean crystal—. In the poplar grove a spring recites its song to the plants.] (103)

The sense of place and detail initially seem to gesture toward the tradition of loco-descriptive poetry. Yet it is an animated, somewhat enchanted nature that is initially represented, and the mystery of its enchantment is related to the personified agency accorded the various natural elements. From the rather abstract but already humanized presence that inheres in the morning stillness—"dulzura infantil"—the poem progresses through a series of subject-verb clusters that specify the constituent components of that initial presence: "un vaho / cubre," "las arañas tienden," "un manantial recita." That these features of the landscape are indeed silent actors is underscored both by the shift from the initial "hay" to transitive constructions, and by the increasingly prosopopeic tenor of the verbs: "cubre," "tienden," "recita." From the outset then, the poem appears to fuse the sensual specificity of the loco-descriptive with an animistic understanding of the natural world. In this regard, the poem's figural landscape is perhaps less distant from romantic and *modernista* pantheism than is initially apparent.

Against the backdrop of this resemblance, however, important distinctions come into relief, and one only need recall the characteristically spiritualized natural world of poets such as Gustave Adolfo Bécquer or Juan Ramón Jiménez in order to appreciate the differences that characterize the present case. Whereas the earlier writers tended to conceive of an ultimate meaning or grand signified contained within nature—that is, the Becquerian "espíritu sin nombre," Jiménez's *Dios deseado y deseante*—the natural world in "Los encuentros" (and in the *Libro de poemas* more generally) does not inevitably lead to the transcendent. There is, to be sure, a kind of spiritualized residue clinging to the natural world, but this tenuous presence is characterized as an illusory and, at best, fleeting "childish sweetness." At the same time, the fragile illusion of a natural world imbued with latent spirit is not the poem's conclusion, but rather a point of departure. Indeed, the initial landscape itself already contains the suggestion of roads that might lead elsewhere in the image of spiders' silk paths etched into a crystaline sky. To phrase it another way, the problem of transcendence, which the poem will take up explicitly, is already at work in the initial figuration of the landscape—the personified agency of the natural—and it will in effect catalyze the ensuing voyage.[4]

It is the contemplation of this quasi-magical landscape that prompts the protagonist of the poem to action: "Y el caracol, pacífico / burgués de la vereda, / ignorado y humilde, / el paisaje contempla. / La divina quietud / de la Naturaleza / le dio valor y fe, / y olvidando las penas / de su hogar, deseó / ver el fin de la senda. / Echó a andar" [And the snail, a peaceful path-dweller, unknown and humble, contemplates the landscape. Nature's divine stillness gave the snail courage and faith, and forgetting the burdens of its home, it desired to see the end of the path. It began to walk] (103). Contemplation generates the desire to see further. Significantly, this desire is engendered by the very illusion of the figurative "beyond" that seems to inhere in the landscape: the "dulzura infantil / en la mañana quieta" [childish sweetness in the still morning] of the first verses now reappears, somewhat modified, as "la divina quietud / de la Naturaleza" [Nature's divine stillness] (103).

Moreover, this divine stillness not only prompts the quest, but also arms the lowly hero—"ignorado y humilde"—with the chivalric virtues—"valor y fe"—that, along with a bit of necessary forgetting, will enable this knight errant to set out on his adventures. Nature's poetic mystery is Dulcinea to a new quixotic hero. To the extent that the romantic project has often been described as the internalization of the romance paradigm, these echoes of quest-romance remit once again to the neoromantic underpinnings of the poem.[5] But once more, the distance that separates Lorca from this legacy is equally clear. In place of the titanic grandeur of the romantic quest hero, the poem charts the odyssey of a snail. This diminished stature can be read as analogous to the new, more muddled role of the transcendent as outlined above: as the quest object is problematized, so is the hero who pursues it.

At the same time, the markedly unheroic, comfortable "bourgeois" qualities of the snail would seem to point away from the exceptionalism of romantic subjectivity to a more common, everyday figure. This more broadly emblematic quality of the poem's protagonist is underscored by the subtle identification between reader and snail that is suggested in the opening sequence. The initial verses structure readerly experience so as to mimic the snail's interaction with the landscape. The first twelve verses—the representation of the landscape—progress without reference to the snail. And even the initial four-verse description—"Y el caracol, pacífico / burgués de la vereda, / ignorado y humilde, / el paisaje contempla"—does not specify the snail's relationship to the landscape until the last possible moment. It is only with the verb "contempla"—significantly deferred by the hyperbaton—that the connection between reader and snail becomes clear. Both have been engaging in the same activity (although at different phenomenological levels). The identification is doubly revealing, for if it tacitly suggests that the snail's quest is in some sense the reader's, it simultaneously reframes the snail's contemplation and, indeed, the question of transcendence, in terms of reading. To the extent that this analogy echoes the conceit that equates readerly experience with the lyric subject's temporal progression throughout the book, the initial sequence of the poem stages one of the *Libro de poema*'s organizational principles.

As the text progresses from its idyllic opening to the two subsequent "encuentros" that constitute the snail's adventure per se, the implicit questions concerning the "beyond" of his journey—whether conceived of as pantheistic mystery, chivalric ideal, the enigma of figuration, or now as the readerly problem of meaning—begin to be addressed in more explicit terms. In the first encounter, for example, the snail overhears a dialogue between two frogs concerning the relationship of song to divine compassion:

> "Esos cantos modernos"
> —murmuraba una de ellas—
> "son inútiles," "Todos
> amiga"—le contesta
> la otra rana, que estaba
> herida y casi ciega—.
> "Cuando joven creía
> que si al fin Dios oyera
> nuestro canto, tendría
> compasión. Y mi ciencia,
> pues ya he vivido mucho,
> hace que no lo crea.
> Yo ya no canto más". . .
> Las dos ranas se quejan
> Pidiendo limosna
> A una ranita nueva
> Que pasa presumida
> Apartando las hierbas.

> ["Those modern songs are useless," one of them murmured. "All of them," answers the other frog, which was wounded and almost blind. "When I was young I thought that if God were finally to hear our song, he would show us compassion. But my science—for I have lived a long time—makes me disbelieve. I no longer sing . . ." The two frogs complain while begging from a young little frog that presumptuously passes by, pushing plants to the side.] (104)

To ponder the futility of song, the absence of divine compassion, and a loss of faith would seem to be the initial consequences of the snail's journey into the world, a journey, it is worth recalling, that ironically was motivated precisely by the snail's desire to see beyond: "deseó / ver el fin de la senda" [it desired to see the end of the path] (103). To the extent that the frogs' "Dios" can be understood more generally as a sign of the transcendent, the first encounter raises the disturbing prospect with which the snail will contend throughout the rest of the poem, the possibility that there is not a "beyond" to be seen, or that at the very least it is inaccessible. Significantly, this notion is linked to the snail's transition from a mythic nature at the beginning of the poem to the world of culture and history as figured by the frogs. The transition is evident in the new lexicon of utility and science that appears in these verses; it is also suggested by the rhetoric of past and present—"creía," "no lo crea"—by the generational contrasts and rivalries—"dos ranas viejas," "una ranita nueva"—and by the social character of the frog-world evoked by the dialogical structure of the passage.

Curiously, the progression from perception to action, which characterized the first moment of the poem, will in this encounter repeat itself in the transition from overhearing the frogs to interacting with them. But in place of the beguiling serenity of nature, which initially prompted the snail to action, the transition is now a fearful, foreboding one that produces horrified silence. "Ante el bosque sombrío / el caracol se aterra. / Quiere gritar. No puede. / Las ranas se le acercan" [The snail is terrified in the face of the dark forest. It wants to shout but cannot. The frogs approach] (104). The terror seems to stem from the metaphysical doubt which the snail has overheard, and its inability to scream is telling in its ironic replication of the very phenomenon that has caused it—the old frog's "ya no canto." At the same time, it is worth noting that the problem of language evident in the poem's initial sequence reappears here in the form of song.

Expression, song, and prayer are highlighted as markers of belief in the frogs' mini-inquisition concerning the afterlife: " '¿No cantas nunca?' 'No canto,'/ dice el caracol. '¿Ni rezas?' / '¿Ni crees en la vida eterna?' " ["Don't you ever sing?" "I don't sing," says the frog. "Don't you pray? Don't you believe in eternal life?"] (104). The snail replies with a question—"¿qué es eso?" [what is that?]—which leads to divergent accounts of the hereafter. For the frogs, the eternal life is "vivir siempre / en el agua más serena, / junto a una tierra florida / que a un rico manjar sustenta" [to live always in the most serene water, next to a flowery land that provides rich sustenance] (105). The snail's account, also species-specific, differs in its terms:

> "Cuando niño a mí me dijo
> un día mi pobre abuela
> que al morirme yo iría
> sobre las hojas más tiernas
> de los árboles más altos."

["As a child, my poor grandmother told me one day that when I died I would go over the tenderest leaves of the tallest trees."] (105)

From the initial communion with its natural surroundings, the snail has thus fallen into the fractured world of social intercourse, a kind of Tower of Babel that juxtaposes competing representations of the transcendent. The immediacy of the figurative "beyond" that was initially inhering in the landscape has given way to an irreconcilably mediated model. And it would seem that there can be no return to innocence. Indeed, the frogs' dogmatic response to the snail as the encounter concludes underscores the dangers—and latent violence—of attempting such a return:

> "Una hereje es tu abuela.
> La verdad te la decimos
> nosotras. Creerás en ella,"
> dicen las ranas furiosas.
> "¿Por qué quise ver la senda?"
> —gime el caracol—. "Sí, creo
> por siempre en la vida eterna
> que predicáis . . ." Las ranas,
> pensativas, se alejan,
> y el caracol, asustado,
> se va perdiendo en la selva.

["Your grandmother is a heretic. We're the ones who tell the truth. You will believe in it," say the furious frogs. "Why did I want to see the path?"—moans the snail—"Yes, I believe forever in the eternal life you preach . . ." The frogs move away, thinking, and the frightened snail slowly wanders off into the jungle.] (105)

The ironies that accumulate in this passage are not difficult to discern: the snail's curiosity in the beyond has engendered a forced declaration of faith, adventure has begotten regret, the purposeful quest progression has led to loss of direction, and an ingenuous search for meaning has come to the dead end of orthodoxy. As a final twist, once the snail flees from the encounter, its inquisitors, who earlier doubted that their songs could inspire divine compassion, are now revealed to lack the very faith they have demanded of their victim: " '¿Crees tú en la vida eterna?' " ["Do you believe in eternal life?"] asks one of the frogs. " 'Yo no,' dice muy triste / la rana herida y ciega. / "Por qué hemos dicho, entonces, / al caracol que crea? / . . . / Me lleno de emoción / al sentir la firmeza / con que llaman mis hijos / a Dios desde la acequia.' " ["Not I," the wounded, blind frog answers. "Then why have we told the snail to believe? . . . (Because) I am filled with emotion when I hear the firmness with which my children call God from the ditch"] (106). Behind this last irony persists the aforementioned preoccupation with language: just as overhearing the frogs' skepticism seemed to accost the snail's sensibilities, so the songs of the younger generation now offer a partial palliative to the frogs' doubt. Common to both is the conception of language as a potential register for vicarious experience.

In the second encounter a more thoroughly secular orthodoxy will clash with glimpses of the beyond as the snail stumbles upon a martyrdom in progress: "El pobre caracol / vuelve atrás. / . . . / Con un grupo de hormigas / encarnadas se encuentra. / Van muy alborotadas / arrastrando tras de ellas / a otra hormiga que tiene / tronchadas las antenas" [The poor snail turns back . . . and encounters a group of blood-red ants. They move riotously, dragging behind them another ant whose antennae have been cut off] (106). The bruised victim explains the reason for its persecution—"la hormiga, medio muerta / dice muy tristemente: / 'Yo he visto las estrellas' " [the half-dead ant says sadly: "I've seen the stars"] (106). The subsequent conversation is structured so as to parallel vaguely the earlier exchange with the frogs. The sequence begins, once again, with a question: " '¿Qué son estrellas?' dicen / las hormigas inquietas. / Y el caracol pregunta / pensativo: '¿Estrellas?' " ["What are stars?" the ants ask impatiently. And the snail asks, thinking, "Stars?"] (107). In this case, however, the ant's response—" 'Son luces que llevamos / sobre nuestras cabezas' " ["They are lights that we have above our heads"] (107)—does not lead to discrepant versions of celestial vision, as was the case with the frogs, but rather to a more direct confrontation with nonvision:

> "Nosotras no las vemos,"
> las hormigas comentan.
> Y el caracol: "Mi vista
> sólo alcanza a las hierbas."
> Las hormigas exclaman
> moviendo sus antenas:
> "Te mataremos; eres
> perezosa y perversa.
> El trabajo es tu ley."

["We do not see them," the ants comment, and the snail says, "My sight only reaches as far as the plants." The ants exclaim, while moving their antennae, "We'll kill you; you're lazy and perverse. Work is your law."] (107)

The disturbing possibility first suggested in the encounter with the frogs now takes center stage. It is no longer the content of the transcendent gaze that is in question but rather the activity itself, and the latent violence that accompanied the frogs' demands has now come to the surface. Corresponding to this new state of affairs is the more explicitly social nature of the conflict. If the first encounter represented the snail's shift from nature to culture, the new interaction with the ants—the social insect par excellence—now frames the conflict in terms of conformity to the demands of collective life. At the same time, yet another romantic topos—the lonely visionary accosted by the insensitive, myopic masses—has come into sight. As was the case with earlier such echoes, however, this evocation of the romantic paradigm ought not to be understood merely as neoromantic affirmation, for it also signals the distance that separates the rhetoric of the *Libro de poemas* from its romantic cousins.

Indeed, a subtle shift has taken place in the second encounter. The conflict no longer bears on the poem's protagonist directly but is instead a scene to which the snail is witness. " 'Contadme lo que ha hecho' " ["Tell me what it has done"], the snail comments with regard to the renegade ant. " 'Yo juzgaré en conciencia' " ["I will judge (the case) in conscience"] (106). For the snail, direct experience cedes to observation once again, inverting its earlier progression from contemplation to action. In this regard, the snail seems to be circling back to its initial contemplative mode. Within this circular movement, however, the poem has effected a series of displacements such that the question of transcendence that informed the opening sequence is now externalized in a very different sense. The transcendent is no longer the enigma of figuration inhering in the representation of nature, as it was at the beginning of the poem. It is now a spectacle of violent contestation and conflict. The disturbing "death of transcendence" which the snail overheard in the encounter with the frogs is now literalized as the death of the romantic visionary represented by the ant:

Por el aire dulzón
ha cruzado una abeja.
La hormiga, agonizando,
huele la tarde inmensa,
y dice: "Es la que viene
a llevarme a una estrella."
Las demás hormiguitas
huyen al verla muerta.

[A bee flew across the sweet air. Dying, the ant smells the immense afternoon and says, "It's the one who's coming to take me to a star." The rest of the ants flee when they see (the victim) die.] (107–8)

Strikingly, the observational stance that the snail assumes is markedly detached in comparison to its blissful, initial contemplation of the landscape. With regard to the question of transcendence, the poem's protagonist can neither be identified with the worker-ants' blindness nor with the visionary ant's insight. Instead, the snail has foregrounded the epistemological limits within which it operates: "mi vista / sólo alcanza a las hierbas" [my sight only reaches as far as the plants] (107). The snail thus hovers between its initial desire to pursue the beyond and its newly discovered recognition of the limits that circumscribe such an endeavor. Such oscillation also seems to have been encoded into the death scene itself. While the ant's death can be read as the end of the romantic visionary, the possibility of the beyond is simultaneously reaffirmed in the dying insect's reading of a passing bee as a possible chariot to the stars. Moreover, the sweetness of the air that the bee traverses—"aire dulzón" echoes the childish sweetness with which the poem opened—"dulzura infantil"—suggesting that the ant's death, like the snail's journey, may in fact be a kind of circular return home. It is, of course, no coincidence that the ant's dying gesture is an act of reading, for as we have seen, the poem has repeatedly framed its questions in relation to language and interpretation.

And although the ant's last words might be read simply as the projection of a pathetic, dying wish, the moment is considerably more complex when one takes into account that this last act in effect mirrors the conceit that readers have been entertaining all along; the ant has simply read the bee as we have been reading snails, frogs, and ants to begin with. The poem thus underscores the basic trope—prosopopeia—that came into relief at the onset of the journey. As such, the reader cannot dismiss the ant's interpretive act without pulling the rug out from under the poetic conventions that have made the poem intelligible in the first place. To the extent that the reader might consequently be described as also hovering between positions, the earlier identification established between the snail's voyage and the reader's is reaffirmed. Indeed, as the poem begins to come to a close the snail in effect "reads" its experiences in an attempt to arrive at some conclusion:

> El caracol suspira
> y aturdido se aleja
> lleno de confusión
> por lo eterno. "La senda
> no tiene fin"—exclama—.
> "Acaso a las estrellas
> se llegue por aquí.
> Pero mi gran torpeza

me impedirá llegar.
No hay que pensar en ellas."

[The snail sighs, and bewildered, it moves off, full of confusion about things eternal. "The path has no end," it exclaims. "Perhaps this is the way to the stars, but my great sluggishness will keep me from getting there. Better not to think about them."] (108)

By linking its own search with the ant's star quest, the poem makes explicit once more the thematic constant—the preoccupation with the transcendent—that has informed the representation of the landscape, the frogs, the ant's vision, and the snail's journey itself. And the sequence of emotions in this passage—confusion, self-doubt, resignation, and finally, a kind of renunciation—initially suggests that the quest narrative which has structured the poem has fundamentally been a narrative of disenchantment. In this respect, it is not difficult to appreciate the ways in which, once again, the ghosts of romantic idealism, particularly the topos of the pursuit of the impossible, seem to haunt Lorca's imagination throughout the poem.[6] Yet in what is perhaps one of the poem's most interesting rhetorical moves, the romantic disenchantment plot is represented as a circular rather than a linear voyage. The closing sequence of the poem remits to its beginning:

Todo estaba brumoso
de sol débil y niebla.
Campanarios lejanos
llaman gente a la iglesia,
y el caracol, pacífico
burgués de la vereda,
aturdido e inquieto
el paisaje contempla.

[Everything was misty, from a weak sun and fog. Bell towers in the distance call people to church, and the snail, a peaceful path-dweller, bewildered and uneasy, contemplates the landscape.] (108)

The approaching evening is, of course, the symmetrical counterpoint to the snail's morning departure, and the final verses bring the poem's cyclical structure into full relief by repeating the words that introduced the snail at the beginning of the composition. It is an imperfect symmetry, however, for this is a return that has been mediated by the poem itself, or more generally, by the passage of time. The dusk haze—"de sol débil y niebla"—contrasts with the earlier morning air—"cristal limpio de aire"—and "aturdido e inquieto," now takes the place of the initial "ignorado y humilde." Moreover, an entirely new realm of experience has made its appearance in the form of the bell towers that call, not to a familiar animal world but rather to the world of humans. The poem thus expands its various meditations on the transcendent to include the historical, human institution of the church. In short, the return enacted here is a return with a difference. The final repetition of the key term, "contempla," by which the snail's journey had initially been linked to the reader's own activity, is exemplary in this regard. For by remitting to the contemplation that preceded the snail's voyage, the repetition of the term at the end of the poem underscores the temporal distance that separates that first moment from the second. It is now a traversed landscape that the snail contemplates, just as the reader now contemplates a completed poem.

At this point it will be useful to take a cue from the poem's structure in order to circle back to the question posed at the beginning of this essay, the question concerning the relationship between "Los encuentros de un caracol aventurero" and the *Libro de poemas.* If, as I suggest, the snail's journey is, among other things, a figure for the "journey" through the collection of poems, and if such movement through the collection is in turn meant to replicate the poetic psychobiography of Lorca's adolescence, a number of preliminary conclusions can be drawn. It is not difficult to appreciate, for example, that the poem, like the book, can be read as an allegory of maturation, a quasi-narrative representation of the passage from childhood innocence to the troubled uncertainties of early adulthood. If spiritual autobiography is the frame of reference, it becomes equally clear that both poem and book chart a crisis in faith. The moments of the Christian master narrative—paradise, the Fall, redemption—are present, but the *redemptive* return seems impossible. If such a frame is expanded further by recalling both the move from nature to culture that accompanies this spiritual crisis, and the closing allusion to the church, it also becomes apparent that the poem, like the book, might profitably be read as an allegory of secularization, the encoding of the historical effects of a desacralizing modernity on an earlier religious metaphysics. Indeed the flux between each pole seems to have been replicated in the aforementioned hovering between interpretive paradigms, which characterized both snail and reader's potential understanding of the ant's death.

To the degree that each of these frames—childhood loss, spiritual crisis, and the drama of secularization—were preoccupations of an historically older romantic rhetoric (as well as the romantic seed that *modernista* rhetoric carries with it), it is no surprise that echoes of the romantic surface with such frequency in the poem and throughout Lorca's idealized representations of his youth. Indeed, the presence of such rhetoric highlights what the compartmentalizing fervor of certain historiographies, along with a voguish celebration of the discontinuous can too easily erase: the *longue durée* of Spanish modernity and the dramas it has engendered.[7] To frame the early Lorca in this way is to momentarily estrange him from the contingencies of 1921 in order to read the *Libro de poemas* against a deeper historical time. Against such a backdrop Lorca's engagement with the neoromantic—indeed a rewriting of the romantic—can be understood metapoetically as an exploration of the limits of that particular cultural legacy. Such a reading becomes particularly suggestive when one realizes that the snail's voyage, whether coded as the pursuit of transcendence, as an internalized quest romance, or as the textual search for an ultimate signified, shares with its romantic predecessors the same basic structure of desire, a structure characterized by that telltale process of deferral in which the object—that is, meaning, Eden, God—is always already beyond grasp.[8] As I have tried to make clear throughout the poem, however, by the time Lorca was writing the *Libro de poemas* this very structure had ossified into an *idée recue.* The poem is in this regard not only a repetition of that structure but also a commentary on it; that is, the poem enacts a familiar neoromantic metaphysical crisis, but it simultaneously registers the crisis of the poetic discourse in which such anxiety has traditionally been expressed. In this regard, the snail's interlocutors, who can readily be associated with poetic discourse—the singing of the frogs, the vision of the ant—represent the weakened claims of the romantic on Lorca's present.

The futility of song and the martyrdom of the visionary ant are signs of poetic discourses that are themselves in the process of disappearing. The poem suggests that the long-lived romantic paradigm of the poet as "reader of cosmic mysteries," which ironically was itself an aesthetic secularization of earlier religious mysticisms, is no longer viable. In short, the snail's search in "Los encuentros de un caracol aventurero" and the lyric subject's search throughout the *Libro de poemas* is, among other things, a search for new expressive modes, a search to which Lorca's subsequent poetic evolution would bear eloquent witness. In this context, the snail's circular itinerary is an especially suggestive metaphor, for it encapsulates with surprising economy the historical dialectic—the return with a difference—that necessarily characterizes both Lorca's poetry, and more generally, the history of all cultural production. The snail's circular movement is a reminder that, as Francis Mulhern has written, the historical is "necessarily complex in character, never belonging to a single mode (continuity/discontinuity) or temporality."[9]

At the same time, as a figure of desire, the snail's circling is an apt image of a libidinal structure destined to repeat itself, despite the new terms in which that structure may be represented. In the case of the *Libro de poemas*, the snail's circling augurs the circular itinerary of the collection itself, as a brief comparison between "Los encuentros" and the book's last poem, "El macho cabrío" (The he-goat), reveals. Although radically different in surface detail, this poem circles back to the theme of desire that was evident in "Los encuentros." In this case, however, the transcendent nature of the quest object, the childhood problem of the early poems, has disappeared. To be more precise, it has been rewritten. In "El macho cabrío" desire is finally expressed in the more modern idiom of Eros.

The lyric subject's enthusiastic greeting of a symbolically charged he-goat could not be, in this sense, more suggestive: " '¡Salve, demonio mudo! / Eres el más / intenso animal. / Místico eterno / del infierno / carnal . . . / ¡Cuántos encantos / tiene tu barba, / tu frente ancha, / rudo Don Juan! / ¡Qué gran acento el de tu mirada / mefistofélica / y pasional!' " ["Hail, mute demon! You are the most intense animal. Eternal mystic of the carnal inferno . . . How many charms are in your beard, your wide forehead, coarse Don Juan! What a great thrust is in your devilish, impassioned gaze!"] (219). The erotic has displaced the metaphysical, in a shift that was anticipated in an earlier poem, "Sueño" (Dream), where the lyric subject had the following vision: "Iba yo montado sobre / un macho cabrío, / El abuelo me habló / y me dijo: / —'Ese es tu camino' " [I was riding a he-goat. My grandfather spoke to me and said, "That is your path"] (205). The image of the path, which was so central to "Los encuentros," reappears at the very moment the grandfather identifies a new, eroticized path for his grandson.

The journey of the *Libro de poemas* is in this sense a progressive unveiling of the sublimated desire at work in "Los encuentros." At the level of the individual, the shift correlates with the sexual awakening of adolescence; as a representation of a young poet's journey, it represents a break with neoromantic idealization; and in terms of a broader cultural history, it can be read as the waning grasp of the sacred on an increasingly modern society. In the end, however, the circular itinerary of both poem and book is a reminder that these very results—adolescent maturation, poetic development, and secularization—are themselves but beginnings: the final poem is, after all, a *greeting*. As one circle closes another opens, which suggests that perhaps it is not the circle but a spiral that in the end best accounts for the trajectories we have described, the slow-moving spiral of history and of desire, the spiral of a snail's shell.

Notes

1. Federico García Lorca, *Obras completas* (Madrid: Aguilar, 1961), 99; all quotes from Lorca's work are taken from this edition and will be cited in the text by page number only. Unless otherwise indicated, translations from the Spanish are my own.

2. See, for example, Candelas Newton, *Lorca: "Libro de poemas" o las aventuras de una búsqueda* (Salamanca: Universidad de Salamanca, 1986), and Ian Gibson, "Lorca's 'Balada triste': Children's Songs and the Theme of Sexual Disharmony in *Libro de poemas*," *Bulletin of Hispanic Studies* 46 (1969): 21–38.

3. Perhaps the most compelling such project to highlight the central role of temporality in romantic rhetoric has been Paul de Man, "The Rhetoric of Temporality," in *Blindness and Insight: Essays in the Rhetoric of Contemporary Criticism* (Minneapolis: University of Minnesota Press, 1971), 187–228.

4. In this regard, early in the poem, the snail's search is linked to the enigma of figuration, specifically, metaphorical figuration. As will become clear later, the oscillation between the literal and the figurative—what Paul Ricoeur calls the "is" and "is not" of metaphor—will have considerable consequences as the poem progresses. See Paul Ricoeur, "Metaphor and Reference," *The Rule of Metaphor* (Toronto: University of Toronto Press, 1977), 216–56.

5. Harold Bloom, "The Internalization of Quest-Romance," in *Romanticism and Consciousness* (New York: Norton, 1970), 3–24.

6. For the nineteenth-century genealogy of this particular brand of idealism, see Michael Iarocci, *Enrique Gil y la genealogía de la lírica moderna* (Newark, N.J.: Juan de la Cuesta, 1999).

7. A recent example of work that focuses Spanish romanticism as an essentially discontinuous process is Philip Silver, *Ruin and Restitution: Reinterpreting Romanticism in Spain* (Nashville, Tenn.: Vanderbilt University Press, 1997).

8. My interest here is not to develop a full-blown psychoanalytic theory of desire but rather to call attention to its structural function in the poem, and more generally, in the *Libro de poemas*. For a much more extensive treatment of desire and its relation to language, see Jacques Lacan, *Écrits: A Selection*, trans. Alan Sheridan (London: Travistock, 1977).

9. Francis Mulhern, ed., *Contemporary Marxist Literary Criticism* (London: Longman, 1992), 22.

Missing in Action: Invisible Males in *La casa de Bernarda Alba*

Robert Lima
The Pennsylvania State University

THE theater of Federico García Lorca has been approached from many directions, so rich in topoi and imagery is its expression.[1] Among the most evident concerns in many of his plays is the relationship between men and women, which is seen to result either in the death of the body or in that of the spirit; in some plays both occur in what I have termed "blood spilt and unspilt,"[2] with most men dying and the women in their circle remaining in life to mourn their passing.

Lorca had a deep-seated affiliation with women from his earliest upbringing, and he sets out to uphold their cause in many of his serious plays, from *Mariana Pineda* and *Doña Rosita la soltera* (Doña Rosita the spinster), to the three rural tragedies that marked his final years as a playwright and as a man. In these works women are the focal point of the action, be they active participants or passive figures in the evolution of the circumstances of their lives. He identifies with what he sees as their plight, the result of society's curtailment of women's rights on many levels.

As to men, Lorca's stance is more ambivalent. Perhaps his attitude is founded on a personal value system that perceived the great inequity of a society that traditionally has given higher status and freedom of expression to men than to women. Furthermore, it could be argued that his point of view stemmed from displeasure over society's acceptance of men in heterosexual bondings, often with attendant infidelities and selfishness, and its rejection of those with alternate lifestyles. Early on, Lorca became aware of his own homosexuality and felt uncomfortable with the knowledge that he was "different"—a very difficult state of life in a Spain that, even in the rapidly changing mores of turn-of-the-century Europe, retained traditional codes of moral and social conduct[3] and condemned homosexual unions. But Lorca did not express himself directly on the subject and so conjecture on the rationale for the bias in his plays against heterosexual males who abused their special status in society must be based upon the demonstrated evidence. Whatever the cause, Lorca shows an evolving negativity toward male characters when presented in male-female interaction; as the ensuing discussion posits, this negativity leads to the exclusion of men from the stage.

If the male protagonist of *La zapatera prodigiosa* (The shoemaker's prodigious wife) appears to triumph over his fickle and shrewish wife, it is only to retreat from his earlier show of manliness into a peaceful state of being, albeit one in which she will dictate the rules of their household. If in *El amor de don Perlimplín con Belisa en su jardín* (The love of Don Perlimplín for Belisa in the garden) Don Perlimplín dominates the action, it is at an extreme personal cost, for he abandons life when he commits suicide in order to awaken his nubile wife to selfless love. In each case, the man becomes secondary, only a means to an end: the makeover of the wife. But in *Así que pasen cinco años* (When five years pass), a surrealistic-symbolic assessment of the mind of El Joven, there is only one player; all other characters are projections of his personality, cast as male and female in order to embody the psychoanalytic elements of his being.[4] Yet, even if the play can be interpreted as male-dominated, it is the failure of his relationship with the female, La Novia, that forms the crux of the protagonist's dilemma: the search for fruition through a process akin to the Freudian reintegration of the personality. Failing in this, El Joven dies.

More so than Don Perlimplín's suicide for love, the surrealistic death of El Joven in *Así que pasen cinco años* signals a new direction in Lorca's dramaturgy, for not only is the protagonist's death unsought by him, but, pointedly, this play is the chronological antecedent of the three tragedies, wherein all male protagonists are made to die or are kept outside the action. With *Así que pasen cinco años* the playwright's negativity toward male characters becomes evident both in their elimination as protagonists and in their destruction or departure from the scene. Thus, in the first two of the three rural tragedies—*Bodas de sangre* [Blood wedding] and *Yerma*—males are no longer the principal players. Their presence is imperative in the plots, for without them there would be no dramatic action, but they are disposed of, in most cases through physical violence on stage:[5] in *Bodas de sangre,* Leonardo and El Novio fight a duel with knives and both perish; in *Yerma,* Víctor soon departs the scene and Juan dies a violent death—strangled at the hands of his wife.[6] In both plots the female characters—bride, mother, wives—are left without their men: the lover and the bridegroom of La Novia, the son of La Madre, the errant husband of La Mujer, the ideal man (Víctor) Yerma might have married and the man she did, whom she immolated—to bemoan their reduced personal and social states. The function of these men over, they are discarded.

Where the plots of *La zapatera prodigiosa, El amor de don Perlimplín con Belisa en su jardín, Así que pasen cinco años, Bodas de sangre, Yerma,* and *Doña Rosita la soltera* are predicated on the actions of men and their involvement with women, in the third of the rural tragedies, *La casa de Bernarda Alba* (The house of Bernarda Alba), the pattern for male death foregrounded in earlier works devolves into the extreme, for now males are missing in the action of the play altogether. Considering the fact that this was the last of Lorca's plays to be completed before his untimely death, this lacuna makes a suggestive argument for interpreting the playwright's direction as one of abandonment of the heterosexual male character in future plays. But this must forever remain a moot point.[7]

In the listing of the characters in *La casa de Bernarda Alba,* the reader sees no males. The lines that follow immediately state: "The poet gives notice that these three acts are intended to be a photographic documentary."[8] This notification alerts the reader to Lorca's bow to reality in the portrayal of Bernarda Alba's household, and it might be conjectured that a household without men was what appealed to the playwright in his theatrical documentation of a real Andalusian (read Spanish) family.[9]

The first man missing in the action of the play is Don Antonio María Benavides, whose death is the motivating factor in the plot of the tragedy. Having died prior to the events that unfold on stage, he is the first invisible male in the context of the play, although there are passing references to Bernarda's father and grandfather, both long-deceased. The dead husband—father-master's "presence" prevails early on. First, it is said of one of his daughters that "during the first responsory Magdalena fainted" (1439), she being "the one who is most alone" because "she was the only one who loved her father" (1440). Second, there is sympathy for the dead man in the declarations of La Poncia to La Criada: "Her husband now has a well-earned rest" (1441). And third, when alone, La Criada exclaims in hysterical sobs:

> "Damn you, Antonio María Benavides, lying stiffly in your burial suit and full boots. Damn you! You'll never again be able to lift my skirt behind the doors of the corral! . . . I'm the one who loved you most of all your servants . . . And am I to live now that you've left? Am I to live?" (1445)

Hers is both a personal expression of grief at not being the object of her master's lascivious (but welcomed) attentions ever again and the traditional plaint expected of all in a household touched by death. It is also, in a larger sense, the cry that punctuates the demise of the man of the house, who will have been the last male to set foot on the premises. Thereafter, no men will enter the house. And neither here nor later does Bernarda display emotion at the death of her husband, the omission perhaps signifying her awareness of his philandering ways. The only signs of loss are in the fainting of Magdalena and the words of the servants, an expression of feelings which Bernarda immediately prohibits. The episode gives the first indication of the antimale mindset that permeates the play, ultimately causing its tragic conclusion.

In this first act, the men who have returned from the funeral service, conducted by the parish priest (another invisible male, along with the other priests who have attended the funeral), do not enter the deceased's house, now most obviously the domain of his widow, Bernarda Alba; they are made to remain outside, on the patio. The dialogue does not disclose if La Poncia's sons, of whom she speaks on occasion, are among the men, thus they must also be counted among the missing in action. After telling La Poncia to give the men something to drink, seemingly a small act of charity to those left out in the blazing Andalusian sun, Bernarda orders: "Let them leave the way they came. I don't want them to go through here" (1447). "Here" being the room of the house where she and her daughters are gathered. Notwithstanding, the men introduce themselves into the house indirectly through the gift that La Poncia brings: "The men collected the money in this bag to pay for the Prayers for the Dead" (1450). With Christian charity toward the memory of one of their own, the men have made a financial contribution, but it is not clear whether the money will be used as they intended or to the direct benefit of Bernarda and her daughters. Bernarda's aloof and noncommittal response is to order La Poncia to thank them and give them another drink, a negative perception of what it is that men want when sex is not a choice.

And who should know better than Bernarda? She has emerged as the central figure in the society of the household upon the death of her husband, become manlike in dominance and attitude. Not only does she lay down the rules as would a man in similar circumstances, but as a consequence of her empowerment, she decrees that no male will be allowed to invade the premises. As in the animal kingdom, the male leader of a harem of females will deny access thereto to any other male. Bernarda has usurped the role of Don Antonio

María Benavides and will not permit any male entry into the harem she guards, or into the bodies of the women therein.

And no doubt because of the prohibition against them, males are the subject of conversation during the postfuneral reception. A girl among the mourners introduces the name of Pepe el Romano, stating that he was present in the church. Angustias claims to have seen him, but Bernarda replies: "Neither she nor I have seen Pepe," adding that another male had been present: "One who was indeed there was the widower of Darajalí" (1447). But he is derided for sitting too close to a woman, although it is women's attitude that Bernarda chides: "In church, women should not look at men, only at the priest, and only because he wears a skirt." La Poncia mutters under her breath: "You're twisted because you lack a man's heat!" (1448).

The daughters Angustias and Adela nonetheless have been secretly looking at the men from behind the barred window and listening to their salacious conversation. Later, La Poncia tells Bernarda: "They spoke of Paca la Roseta. Last night they tied her husband to a trough and took her on horseback to a high olive grove . . . She didn't seem to mind at all. They say that her breasts were bare and that Maximiliano was holding her as if he were playing a guitar" (1455). The sexual freedom of men (and of the women who welcome their advances) is in contrast with the physical bars that prevent Bernarda's daughters from expressing their own sexual needs. Those bars may have been in place traditionally to keep at bay the outsider—thief or suitor—but they have now become the means of imprisoning the very women they had been erected to protect. Theirs has become the equivalent of a prison or a secular convent—each equally a place for women without men. And their incarceration is to last eight years.[10] When Magdalena objects, Bernarda lashes at her verbally: "I give the orders here. You can't go crying to your father anymore" (1452). Don Antonio María Benavides, although dead, is again resurrected in the dialogue to emphasize the absence of the last man of the household. He will not be there for his favorite daughter, indeed for any of them, ever again. The finality is evident in the situation in which they find themselves upon his death. Bernarda now dictates the gender-specific roles that she perceives her society to assign. "Needle and thread for the female. A whip and a mule for the male" (1452): the women inside the house at sewing chores in preparation of the wedding trousseau, the men outdoors tilling the land into fruition.

Ironically, neither the sewing nor the fruition that normally follows marriage will obtain in the case of Bernarda's daughters by the deceased, for none has had a suitor to speak of (although Enrique Humanas would have courted Martirio had Bernarda not turned him away as unsuitable). Discussing the murderous and abusive behavior of another woman's father, Martirio, bitterly echoing what her name implies, says: "Men cover up for each other," adding "none of them are willing to tell on the other" and "it's better never to see a man. I've been afraid of them since I was a girl" (1459). And it may be that her fear of him and his fear of Bernarda combined to scare away her only suitor. Pepe el Romano, however, has been courting Angustias, Bernarda's eldest and the issue of an earlier marriage, which allows for her marriage despite the mourning. Pepe, at twenty-five and handsome, and Angustias, at thirty-nine and ugly, make a disparate couple but what he seeks is what Martirio expounds on as typical of males: "What do they care about beauty! What matters to them is land, oxen to till it, and a submissive bitch who will feed them" (1460). Angustias, who will bring a sizable dowry from the wealth she has inherited from her stepfather to the detriment of his real daughters, will suit Pepe el Romano very nicely. But Adela learns from her sisters of Angustias's preparations for her marriage to Pepe el Romano and reacts in a manner that prefigures her admissions late in the play. It will become clear that Pepe el Romano, no matter Martirio's assessment of what men want, indeed has an eye for beauty.

Lorca's treatment of men in the first act ranges from Bernarda's perception of them as easily satisfied by drink, to La Poncia's story of their whoring with Paca la Roseta, to Martirio's perception of them as an egotistical lot. However, there are instances of other invisible males in the action of the play's first act who do not fit any of the three categories: when Don Arturo arrives to arrange the distribution of the inheritance, he is not allowed to enter the house; the vendor of the embroidery for Angustias's wedding is likewise kept outside the house; nor has the handsome young doctor who is treating Martirio allowed entry; Prudencia's husband is talked about but, like all the other men referred to, he never appears on stage. Lorca's mission of omitting men from the stage has been promoted most strongly from the very inception of the play's opening act to its closing moments when Bernarda's elderly mother, María Josefa,[11] taken by the household to be mad and kept shut in, shouts out rebelliously: "I want to get married to a gorgeous male" (1470), thus verbalizing the deepest desire of each of her younger co-prisoners. Moments earlier, however, María Josefa had spoken words to her granddaughters that augur what is to come: "Not one of you will get married. Not one!" (1470).

In the opening moments of the second act, the topic of the discourse is the ongoing courtship of Angustias by Pepe el Romano at her window and his straightforward proposal of marriage: "You know that I've been after you; I need a wife who is good, temperate, and that is you if you give me your consent" (1475). As if to highlight his unromantic, matter-of-fact attitude, a discrepancy arises regarding the time of his departure the previous night: Angustias says he left at one-thirty in the morning, but La Poncia heard him depart at four. What happened during the extra hours remains a mystery, one that casts a shadow on Pepe's relationship with Angustias. Could it be that he has been pursuing another woman in Bernarda's household, already acting the despicable male even before marriage? Again to foreground Pepe's cold proposal of marriage, it is contrasted to the impassioned words and gestures of La Poncia's husband, as she discloses to the young women gathered around her: "It was very dark . . . Evaristo came closer, so close that he was trying to get through the bars, saying in a very low voice: 'Come, let me feel you!' " (1476). But that passion did not last long, as La Poncia reveals to the inexperienced women who hang on her every word: "men, two weeks after the wedding, substitute the bed with the dining table and then the dining table with the tavern" (1476). But La Poncia adds that unlike women who cry their eyes out in those circumstances, she stood up to her husband and even came close to putting out one of his eyes in a fit of anger over his comings and goings. Magdalena revels in that confrontation, wishing that all women could be so strong.

Despite her seeming openness in conversing on sexual matters, La Poncia is prudish when it comes to confronting Adela over what she suspects is the young woman's illicit involvement with Pepe el Romano. She has seen her standing nearly naked at her window during Pepe's visit to Angustias and urges Adela to forget him, while at the same time promoting hope by projecting that Angustias will die in childbirth: "Then, Pepe will do what all widowers do in these parts: he'll marry the youngest, the most beautiful of her sisters and that will be you" (1481). But Adela's passion is not to be postponed; what must be must be, she rants, for "looking into his eyes I seem to be drinking his blood slowly" (1482).[12] La Poncia is unable to reason with her desperately impassioned charge and abandons her cause to resume telling suggestive tales about men when Adela's sisters return:

> "The reapers came early yesterday morning. Forty or fifty good-looking young men . . . Last night a dancer dressed in sequins and playing an

> accordion came to town and fifteen of them made a deal with her to go to the olive grove . . . Years ago another of these women came here and I myself gave my elder son the money to go with her. Men need such things." (1485)

La Poncia not only condones such actions by males, she abets them. Yet she cannot abide such libidinous expression among decent women. Adela and her ilk are not free to act like the woman who sold her body. As if to underscore her imprisonment, the chorus of the reapers is heard offstage in a song suggestive of sexual union. But while their lyrics penetrate into the household of repressed females, these men, too, pass on by, their licentious song fading in the distance as Adela and her sisters peer longingly at them through a shuttered window. It is only moments such as this that Bernarda's daughters can look forward to—hearing lusty tales of men told by La Poncia, listening to lusty male voices raised in song out in the street, envying men their freedom. But listening and looking are the only recourse for the trapped women, for the men are missing in action, invisible on the stage of the play and in their lives.

And then the missing photograph of Pepe el Romano—stolen from Angustias's bed—instigates a quarrel until Martirio is found to be the guilty party, not Adela as La Poncia suspected. Nonetheless, knowing what she knows, La Poncia tries to convince Bernarda that there is a better match for Pepe than Angustias: "Adela. She's the one really meant for Pepe el Romano!" (1500). And the story that La Poncia's elder son has told of seeing Pepe near the house at four thirty in the morning, when Angustias claims that he leaves her window at one, implies that he is seeing Adela. There is no doubt also in the eyes of Martirio because she has seen Adela in Pepe's arms. The confrontation that is about to take place between the two sisters is halted by the news that an unwed mother who killed her *male* child is being pursued by a mob of men intent on carrying out vigilante justice on the murdering fornicator.[13] The act ends with Bernarda adding her condemnation and shouting for the fallen woman's execution at the hands of the mob. These events are meant to be a prefiguration of the play's outcome, although it will not be men who bring about the tragic action.

The third and final act commences with Bernarda's conversation with Prudencia, in part about the visitor's husband, who is admired by Bernarda as "a real man" for having ostracized his brothers over an inheritance and his daughter due to her disobedience. Bernarda sees herself in a similar role with her own daughters and servants, who are potentially rebellious against her authority now that the master of the house is dead. No sooner is that association made than another male element is introduced into the dialogue when the breeding stallion starts kicking the walls of his stall. Prudencia ties the two together when she says of the animal: "Contentious as a man" (1509). In a rather evident irony, Bernarda will permit the horse to go to stud at dawn while having committed herself to reining in the sexual drive of her daughters during the extreme period of mourning, in effect rendering them barren for the rest of their lives. Except, of course, for Angustias, whose upcoming nuptials are again the subject of discussion, as is her suitor's announced absence that night.

But will Pepe indeed be traveling with his mother? The earlier question of Adela's sexual misconduct is answered late that night when Martirio confronts her sister, whose hair is uncombed and whose nightgown has traces of straw. The self-serving male, whose proper courtship of Angustias has been at her barred window, has had a more intimate involvement with the youngest of Bernarda's daughters, and Martirio has spied upon their passionate encounters. Pepe has secretly entered the corral on many occasions to make love to Adela. While the man has not set foot in the house proper but only in one of its appurtenances, he has in the course of things therein penetrated Adela, taking her virginity (given quite willingly) and punctured the family's honor (also surrendered by Adela through her willful acts). In her despair at never marrying the man to whom she has given her virginity, Adela has become reckless with her virtue because she knows that her lover will not be hers in marriage. She flaunts her intention of living as the mistress of Pepe el Romano. When Bernarda shoots at the male intruder and acts as if she killed him, Adela commits suicide.[14] The rebelliousness that Bernarda feared has taken a different tack. Adela has died by her own hand for a man she believes to be dead, a Juliet to a Romeo who has not appeared on stage. What began with the death of a male (Don Antonio María de Benavides) ends with that of a female.[15] Death thus encircles the action of the play, like Uroborus chasing its own tail.

The deceased Don Antonio María de Benavides, the male mourners, Bernarda's father and grandfather, La Poncia's husband Evaristo el Colín and their two sons, Don Arturo the Notary, the widower of Darajalí who sat too close to a woman in church, the priests who co-celebrated the funeral mass, the nameless embroidery vendor, the robust young harvesters going to the fields and later singing as they return, Prudencia's husband, Maximiliano and the other buyers of Paca la Roseta's sexual favors, Enrique Humanas who had sought the hand of Martirio, the male baby killed by its mother, the farm hands whom Bernarda employs, and, most notably, Pepe el Romano . . . all are discussed in the play, but those who are alive never appear on stage. They are presences from the male-dominant external society now kept at bay by the despotic matriarch who guards her territory with feral tenacity and keeps control over the women who inhabit it, women unhappy in the realization that males will not again enter a terrain which will be exclusively female.

In the end, Lorca has not only made men invisible in the stage action of the play, he has also denigrated them through sarcastic, often biting words from female characters (Bernarda and Martirio, for two), and through references to immoral or cruel actions on their part that go unpunished. Furthermore, unlike the sacrificial deaths of males in *Bodas de sangre* and *Yerma,* Lorca has deprived Pepe el Romano of a status that would have made him heroic in the eyes of Bernarda's daughters; instead, the playwright has him fleeing the premises upon Bernarda's ill-aimed attempt on his life. Pepe el Romano's cowardly exit indicates yet another mockery of maleness in Lorca's play.

In eliminating males from *La casa de Bernarda Alba,* Lorca has also insured that there will be no possibility of a future male presence in the household. Pepe el Romano's illicit involvement with Adela, Bernarda's attack on him, and Adela's suicide have made Angustias's marriage impossible. What had seemed to be the rantings of María Josefa earlier has now become a true prognostication. Lorca has decreed both physical death and death of the spirit for Bernarda's household; both result from her overly harsh control, which carries Lorca's implicit condemnation as a manlike excess of authority. In years to come, as in the present of the play, men will be absent from the closeted lives of Bernarda and her daughters. Only in their private dreams may a male figure present itself for delectation (as in the mad images in María Josefa's mind) and masturbatory satisfaction, but that will serve only to mask the great frustration that the lack of sexual and maternal fulfillment promises.[16]

Lorca has brought his plays to this momentous state, the result of a process that has systematically belittled the heterosexual male character. He has attacked the male's exalted position in society, his blatant sexual prowess, his disregard of social and moral codes, his cavalier attitude toward women. It may have been Lorca's intention after *La casa de Bernarda Alba* that all males be missing in action in his plays.

Notes

1. In my case, I have analyzed the plays early on in Robert Lima, *The Theatre of García Lorca* (New York: Las Américas, 1963), the first book in English exclusively on his dramatic works. At the time, *El público* had not been published.

2. See Robert Lima, "Blood Spilt and Unspilt: Primal Sacrifice in Lorca's *Bodas de sangre*," *Letras Peninsulares* 8, nos. 2–3 (Fall 1995): 255–59.

3. Lorca's family, particularly through his brother Francisco, its spokesman, has consistently denied his homosexuality.

4. For a fuller discussion, see chapter 7 of Lima, *Theatre of García Lorca.*

5. Unlike Greek tragedians, who kept death offstage, Lorca presses home the dramatic quality of a duel to the death (*Bodas de sangre*) or an impassioned execution (*Yerma*) by presenting the violence of each front and center.

6. See Robert Lima, "Toward the Dionysiac: Pagan Elements and Rites in *Yerma*," *Journal of Dramatic Theory and Criticism* 4, no. 2 (Spring 1990): 63–82.

7. The publication of *El público* may show another direction that Lorca could have taken, but its emphasis on homosexual themes adds potential support to the theory of his abandonment of the heterosexual male character by positing the possibility that Lorca wished to concentrate on gay men in later plays.

8. Federico García Lorca, *La casa de Bernarda Alba,* in *Obras completas* (Madrid: Aguilar, 1971), 1439. Hereafter cited in the text by page number only. All translations from the Spanish are my own.

9. Lorca may have been prompted in his documentary approach by the work of his friend, the filmmaker Luis Buñuel. While in the employ of Spain's Republican government, Buñuel made such documentary films as *Tierra sin pan* (Land without bread), which showed the extremely bad living conditions of people in the medieval villages of Las Hurdes (southwest of Salamanca). Lorca employs real-life people and their names, including Angustias's suitor, as a point of departure for his "photographic documentary."

10. There has been some comment on Lorca's use of numbers in this play, namely the "two hundred women" who enter Bernarda's house as part of the ritual of mourning, and here of the excessive number of years assigned by Bernarda for mourning. What has been taken to be exaggeration was in fact what the playwright intended, thus indicating in the first case the Spanish practice of the entire population of a town joining in the plaint for the deceased, and in the second instance the need to emphasize the extent of Bernarda's fanaticism and the harshness of her dictum.

11. Her name mirrors two of the members of the Holy Family—Mary and Joseph. The third, their child, is omitted, perhaps to indicate the incompleteness of María Josefa's motherhood. Instead of a son who might have sanctified her status and fulfilled her affiliation with the Holy Family, she had Bernarda, who has become her jailer.

12. This statement is reminiscent of Valle Inclán's *Ligazón* (blood pact), in which La Mozuela [The Lass] makes a blood pact with El Afilador [The Knife Sharpener], her intended lover.

13. The episode of the adulteress brings to mind a similar situation in Valle Inclán's *Divinas palabras.* But there Mari Gaila has not killed anyone, much less a newborn. In the context of Lorca's relationship with Valle Inclán, it is interesting to note that in 1936 he participated in the most important act of homage offered upon the death of the famed member of the Generation of 1898.

14. Unlike the duel in *Bodas de sangre* and the execution in *Yerma,* death occurs offstage.

15. Other than Mariana Pineda, who submitted to the gallows rather than betray her lover and friends, Adela is the only woman to actually die in one of Lorca's plays. And, she is the only woman to suffer a physical death in the tragic trilogy. Like Mariana, she has died for love

of a man. Man, then, is held responsible for her demise although it was Bernarda's action that prompted the course of death. It can be said that in keeping with her newfound authority in the household, Bernarda was acting like a man protecting his own.

16. Whether stated or not, extreme sexual and maternal frustration can lead to death in Lorca's plays. *Yerma,* which immediately precedes *La casa de Bernarda Alba* in the trilogy, is a case in point.

Shall the Circle Be Unbroken? Verbal and Visual Poetry in Lorca, Buñuel, and Dalí

Antonio Monegal
University of Pompeu Fabra

THE title of one of Lorca's poems, "Fábula y rueda de los tres amigos," from his book *Poet in New York*, immediately brings to mind the circuit of artistic exchange between these three most famous of friends, Lorca, Buñuel, and Dalí. I do not intend to suggest here that the poem should be interpreted as referring to these three "amigos." I am not going to spend any time trying to defend that interpretation, because I am willing to admit that I am twisting the obscure meaning of the title of the poem to fit the not so obscure objectives of this essay. Just for the purpose of organizing my argument, by "fábula" [fable] I mean the anecdotal evidence we have about the friendship between three of the most remarkable and universal artists born in Spain during this century. By "rueda" [round or wheel] I mean the circuit of exchange of aesthetic ideas among the three artists, their shifting between different art forms, and how their respective practices were affected by it.

The word *fable* does not imply that what we are dealing with is fiction. On the contrary, that part of their common biography has been extensively researched and documented by Ian Gibson, Mario Hernández, Rafael Santos Torroella, and Agustín Sánchez Vidal, who has called his book about the three artists *The Endless Enigma*, after the painting by Dalí.[1] I am calling it a fable because their story is so fascinating and entertaining that, without any addition, it can already make a great novel. They met in the *Residencia de Estudiantes* in Madrid in the early twenties, and the beginning of their careers coincided with a crucial period of Spanish history and a turning point in the history of Western art, a revolution led by the avant-garde movements which reached Spain like a tidal wave. Their fates were different, but in each case determined by the Spanish Civil

War: Lorca was murdered by members of the Francoist forces at the beginning of the uprising; Buñuel went into exile and made most of his films in Mexico and France; and Dalí, after a brief absence, became the only painter of international reputation to accept the protection of the dictatorship. Lorca is probably, in spite of his fame, the lesser known of the three, simply because, as we will all agree, the language of the poet is less universal than those employed by the painter and the filmmaker.

Is it really so? Let's start by playing games with language. If by the language of the poet we mean whether he or she writes in Spanish, English, Russian, or Chinese, we may concede that language is less universal than painting or film (at least silent film). We should take into account, though, that the reading of a painting or a film also requires to bring into play cultural conventions which may be as idiomatic as linguistic ones. Aside from this consideration, if by the language of the poet we mean that which makes his or her use of a language poetic, that which makes poetry what it is, then it is less obvious whether such a phenomenon is universal or culturally bound. The answer will depend on our definition of poetry.

Whatever we think, the case in point here is that, for the three artists we are discussing, poetry is not a matter of language, or, to put it in a different way, the matter of poetry is not necessarily verbal language. A poet is a poet is a poet translates for Lorca, Buñuel, and Dalí into a poet is a painter is a filmmaker. And translation is certainly the issue here, because what is at stake is the possibility of translating, of transporting this ineffable entity called poetry from one art form to another. Lorca, Buñuel, and Dalí believed in that possibility, and it became the center of their aesthetics.

Each of them considered himself a poet; in the first place because each of them wrote poetry. Although Lorca is the one who acquired his reputation as a poet, his friends the painter and the filmmaker also published their poems, mostly in avant-garde magazines, during the twenties and thirties. In addition, Dalí and Buñuel collaborated in two films, *An Andalusian Dog* and *The Golden Age*, and Lorca himself wrote a screenplay, entitled *Trip to the Moon*.[2] He was also an accomplished musician. Buñuel experimented briefly with the plastic arts, in collaboration with Alberto Giacometti to produce a surrealist object, entitled "A Giraffe."[3] And Lorca has left us numerous drawings, which have circulated in many exhibitions and are collected in a catalogue.[4]

So the first manifestation of the "rueda" is the circulation of a single artist between different art forms. This is generally recognized as a case of what we call multiple talent. Many examples come immediately to mind: William Blake's engravings, Pablo Picasso's poetry, to cite a couple. But the concept of the multiple talent is clearly a misnomer, as it suggests an equivalency between the different talents. It is theoretically possible, but historically we can only think of it as an exception. There is little doubt that the majority of those versatile artists, including Lorca, Buñuel, and Dalí, are best known for what they did best. Thus, there is a qualitative hierarchy between their artistic practices. Which does not imply that one art form is subordinate to the other. For any given painter who writes, or writer who paints, it is conceivable that both practices could be absolutely autonomous. It is not enough to know that the author is the same. To conclude that there is a relation, one should find proof in the comparison between specific works. Only then can we decide whether there is some degree of influence or contamination between the different artistic practices, or whether they could even respond to or be ruled by a single aesthetic program.

Lorca certainly did not consider himself anything other than a poet when he was drawing. We have his own words to that effect, in a letter to his friend Sebastià Gasch: "I am now starting to write and to draw poems like the one I am sending you dedicated. When a topic is too long or has a poetically trite emotion, I resolve it with my pencils."[5] Lorca's statement is echoed by Jean Laude in a commentary on Victor Hugo's drawings: "One imagines that [Hugo's concern with the graphic arts] arrives in the nick of time to 'relieve or inspire' the writing, when the written word reaches the periphery of the domain which it would be unsuited to explore."[6]

The second manifestation of the "rueda" refers not to the movement of the artist between art forms but to displacements among the arts themselves, to circuits of exchange and contamination. In the case we are discussing, the circuit is complicated by the fact that we are not limited to the different art forms explored by a single artist, but we are dealing with three artists and, at least, three art forms.

Is there, or better, can there be a common ground between the arts that allows for such an exchange? The issue of the relation between the arts has been the source of many arguments since classical times, and I am not going to try to summarize, much less resolve it, here. But if there is a period that exhibits an unquestionable tendency to this kind of interartistic communion, it is precisely the first half of this century. And right in the middle of that era the relationship between our three friends was at its height. The avant-garde movements, particularly Dada, futurism, and surrealism, generated numerous manifestoes that proclaimed unified aesthetic programs and encouraged the interaction between the arts.

In 1928, Dalí signed one of those programmatic documents, the so-called Catalan Anti-Artistic Manifesto, prepared by a small group of members of the Catalan avant-garde. Lorca probably participated in the drafting of the document, but in the end his name did not appear among the signatures. We can find many reasons for that, but the most obvious is that, in spite of everything the three friends shared, there were serious disagreements in their aesthetic views.

Isn't it surprising that such close interaction between the arts would be proclaimed and actively sought at a time when art forms were becoming more autonomous and self-contained? During the avant-garde period the aim of the different arts was no longer to copy, to imitate reality, but to re-produce it, to produce *artifacts* which are things in themselves. During the time in which both painting and poetry were considered forms of imitation, they had at least in common the world they represented. But when painting is a world in itself, made up of forms and colors, and poetry another world, made up of words, how can we relate them?

Avant-garde artists had at least a common enemy: they all agreed in their rejection of traditional forms of representation, particularly what we could call the humanistic tradition, that interpreted art as a means to communicate ideas, values, and feelings. It is the classical idea of the justification of art as a useful example, in the name of some higher intellectual or moral principle. This is exactly what Dalí and his friends attacked from their anti-artistic position. They were obviously not against art, being artists themselves, but against specific values, such as academicism and sentimentality, which, in their view, infect art and make it rot. For Lorca, Dalí, and Buñuel the concept of putrefaction in art represented old-fashioned, traditional aesthetics, while *poetic* was the adjective they applied to their own art.

We have mentioned how Lorca considered his drawings to be "linear poems." Dalí, in a 1927 article about the public's reaction to one of his first exhibitions, said that they were captivated by the *fet poetic*, the poetic fact contained in the paintings.[7] As late as 1958, Buñuel lectures in Mexico on "The Cinema, Instrument of Poetry."[8] Is there anything of substance behind this apparently indiscriminate uses of the words *poetry, poems, poetic*? Or were these people simply speaking "poetically" and thus we should not take what they said too seriously, too literally? Why bother with such minor detail? It happens to be, at least for me, a very challenging detail, and one that may lead us to a promising approach to the at times confusing discussion of the relation between the arts.

As I mentioned before, there are several excellent studies that analyze the exchanges between the three artists. The repetition of a number of motifs in their respective works has been traced by the Lorca specialists I have already named. Mario Hernández says: "One can prove rather easily that some of Lorca's motifs relate his work to that of Dalí and of other Surrealists, with their tragic atmosphere of organic mutilations, severed heads and hands, and transparent bodies offering a glimpse of a complex circulatory system."[9] But my main interest is not the repetition or the motifs themselves, rather how they function within a given system, how they become *poetic* signs. I will explore a small point: in what way can we say, without being too vague, that all this is poetry, that there is a verbal, literary form of poetry and also visual, cinematic, or pictorial forms of poetry? I would like to stress similarities as well as differences, because the moon, for example, shows up insistently in Lorca's poetry, in Dalí's paintings, and in Buñuel's films, but it also shows up in the sky at night. And it is evidently not the same moon in the sky, in a film, in a painting, and in a text.

We should probably start with the obvious, which is never so, and try to decide what we mean when we refer to something as poetic. It might have to do with the emotion or reaction it provokes. But in that case the poetic nature of a particular work would depend on the sensitivity of the reader or spectator. We tend to confuse poetry with lyricism. When we see an image of an eye being cut by a razor blade, some may say it is poetic, but for many it will simply be disgusting. In that case the definition of poetry would be contingent on the resiliency of the spectator's stomach.

I am going to suggest, as a working definition (without specialized jargon), that poetry has to do with the way in which things (the kind of things we call signs) are put together, and with a way of reading those things, of making sense (or nonsense) of them. Poetry is one among many systems of construction of meaning. It is not merely informative; it does not say everything explicitly. On the contrary, it is made not only of signs, but also of silences. It contains gaps that are to be filled in interaction with the reader or spectator. Meaning is not openly displayed, because within the realm of the poetic the shadow of mystery is always present. The title of the English edition of the catalogue of Lorca's drawings is *Line of Light and Shadow*, and it accurately draws the picture that describes the poetic operation.

Someone may say that, if it is just a matter of mystery, then it is the same as in every form of art, because all of them share that element of mystery. That's precisely why Lorca, Buñuel, and Dalí thought that poetry was the common denominator between the arts. It does not have to be there, it is just a potential. When Buñuel talks about cinema as an instrument of poetry, he is not referring to mainstream Hollywood movies; he is actually attacking them. He is talking about the type of films that take advantage of a quality of the cinema that approximates the texture of dreams. In a way, we can say that there is poetic and prosaic cinema, as there can be prosaic and poetic drawings.

We can thus approach works identified as poetic from the perspective of their enigmatic, or discordant, combination of signs, focusing on the mystery they contain and the challenge they pose to our interpretation.

In a Twilight Zone fashion, there is a text by Dalí that exists in a dimension parallel to a painting: a prose poem, in Catalan, of 1927, entitled "My Female Friend and the Beach," with an epigraph that reads "Honey is Sweeter than Blood," which is the title of a painting of the same year. I cite two fragments:

> Right now, on the beach, the printed letters of the newspaper are eating the stiff, putrid ass . . . My female friend is lying down with her limbs tenderly severed, full of flies and of small aluminum propellers that approach her semi-vegetal nudity, smeared with all the eyeshadow I had given her for her birthday. (*AD*, 49)

These short sequences are enough to perceive the quick jumps between incompatible meanings, the logical discontinuity that governs the text as much as the painting. We can find similar scenarios in Lorca's prose poems of the same period. From "Submerged Swimmer":

> The following morning the countess of X was found on the beach with a liquor fork stuck to the back of her neck. Her death must have been instantaneous. On the sand they found a piece of paper stained with blood that said: "As you are not able to become a dove, you are better off dead."[10]

Death and mutilation serve a double function: they have a specific symbolism and are also employed as mechanisms by which the human body becomes objectified, is converted into a sign. Lorca also resorts to these bloody procedures in other prose poems, such as "The Beheading of the Baptist," "The Beheading of the Innocents," and "Suicide in Alexandria": "When they placed the cut head on the office desk, all the windows in the city broke" (*OC*, 3:156). And in many of his drawings.

One may be stricken by how cruel and horrible all this looks, but in that case we are forgetting a fundamental distinction between art and reality, underscored by Dalí in his response to an art critic who was also scandalized by this kind of paintings:

> the beheaded figures live their organic and perfect life, rest under the shadow of the most bleeding vegetation *without being soiled by blood,* and they even lie naked over the cutting and bristly surfaces of special marbles without risk of death. Will we have to remind Mr. Benet that the *life* of the creatures which inhabit the surfaces of canvases and the world of poetry obey and have life conditions very different from those of the creatures who inhabit the surface of the earth? . . . That the plastic or poetic life of a painting or a poem obeys other laws than those of the circulation of blood? . . . That a beheaded figure, in the plastic or poetic world, is not a figure without a head? (*AD,* 62)

It is not the mutilation itself that is poetic, but the kind of recombination and juxtaposition of signs it allows, a new model of composition that can be compared to the "cut and paste" method of collage. The procedure is usually associated with the visual arts, but we find its verbal equivalent in these disjointed texts. Their mystery or obscurity is the direct effect of what Michael Riffaterre calls "semantic incompatibility," that lack of logical continuity between the elements combined.[11] As a result this poetic construct is not only difficult to interpret but very often impossible to visualize.

The moment we approach the category of the poetic mode from both the verbal and the visual angle, we encounter the concept of what we call, in a blatantly metaphorical sense, the *poetic image.* It is a metaphor because, in principle, the poetic image has little to do with sight. Lorca himself will develop these theoretical distinctions in two of his lectures, "The Poetic Image of don Luis de Góngora" and "Imagination, Inspiration, Evasion." In the first one he states that "all images are displayed in the visual field."[12] But in "Imagination, Inspiration, Evasion" he will move away from that model and during the period when he is writing *Trip to the Moon* and *Poet in New York* he will advocate a poetics which authorizes the production of poetic images that cannot be visualized. He experiments in his writing with verbal constructions he calls "hechos poéticos" [poetic facts or events], which are not at the service of representing reality. He describes it, in a manner close to Vicente Huidobro's *creacionismo,* as a

> "fact" with its own life, new laws, and that [. . .] breaks with any logical control. A poetry full of an order and a harmony that are exclusively poetic. The latest generation of poets is concerned with reducing poetry to the creation of the poetic fact, and with following the laws that it im-

> poses, without listening to the voice of logical reasoning or to the measure of the imagination. They attempt to liberate poetry not only from anecdote but also from the riddle of the image and the planes of reality, which is equivalent to taking poetry to the last level of purity and simplicity.[13]

The poetic mode is thus not an ill-defined spiritual property but a type of verbal construct that suggests a meaning concealed by its mystery, because the words cannot be deciphered in rational and referential terms. From that point of view, only certain forms of assembly of signs qualify as poetic.

We can illustrate this point by analyzing the use of another poetic commonplace, the moon. Naming it, filming it, painting it, can be reduced to prosaic operations. But in the poem that gives title to Lorca's lecture we find the following poetic image (based on the traditional association between the moon and the goddess Diana):

> Diana is hard,
> but sometimes her breasts are banked with clouds.
> (*OC,* 1:451)

If we compare these verses with the initial sequence of *An Andalusian Dog,* by Buñuel and Dalí, it is possible to identify a similar kind of association. The verbal metaphor is echoed in a juxtaposition generated by montage, a purely visual analogy established between two parallel series of shots: clouds and moon, razor and eye, with Buñuel himself acting as the agent whose *vision* establishes the connection. But this association is in fact a result of the spectator's reading. The film takes advantage of the illusion of continuity to make us interpret that the character who looks at the moon is the same one that cuts the eye, when a careful observation allows us to verify, for example, that one is wearing a tie while the other one is not. This is the kind of poetic image in which the elements refer to each other, deriving its meaning exclusively from this interaction. The sequence does not record or represent an actual event, but it produces a figure, a poetic event.

As the film was made before Buñuel and Dalí joined the French Surrealist movement, instead of simply classifying it as an example of surrealist filmmaking, it can be seen as the product of shared concerns and poetic strategies that had already found expression in writing and in painting.

At the textual level, Buñuel first gave the title *The Andalusian Dog* to a book he never got around to publishing and in which we find the prose poem "Ice Palace":

> Near the door there is a hanged man swinging over the abyss surrounded by eternity, howling of space. It's me. It's my skeleton where only my eyes are left. At times they smile, at times they are cross-eyed, at times THEY ARE GOING TO EAT A BREAD CRUMB INSIDE MY BRAIN. The window opens and there is a woman polishing her nails. When they are sharp enough, she pops out my eyes and throws them into the streets.[14]

And in Dalí's "My Female Friend and the Beach" we can read:

> My female friend enjoys the sleepy morbidity of bathrooms and the sweetness of the very thin scalpel cuts on the curved pupil, dilated to extract the cataract. (*AD*, 50)

This obsession with eyes can be linked to a new ideology of seeing, to a break with traditional forms of perception and of representation. And if his friends write about it, Lorca also addresses this subject by visual means, in his drawings and in his screenplay, his own *Trip to the Moon,* where we find several references to eyes and to visual perception. The instructions for filming describe "a frightened head that stares at a point and dissolves into a wire head against a background of water." In a different section there is "a close up of an eye superimposed on a double exposure of fish." And in another, "the moon is cut and out of it appears a drawing of a head that vomits and opens and closes its eyes" (*VL*, 60, 62, 63).

In this strange new gaze of a cinema of poetry we recognize Lorca's response to *An Andalusian Dog.* We know that he wrote *Trip to the Moon* in a couple of days as a result of some conversations about cinema he had with Mexican filmmaker Emilio Amero in New York at the end of 1929 or the beginning of 1930. Amero has recalled that they talked about the release of *An Andalusian Dog* in Paris, in June 1929. Although it is unlikely that Lorca could have had time to see the film in Paris on his way to New York, he probably had enough information for it to have left some traces in *Trip to the Moon* and particularly for the screenplay to be somehow a reply to a film in which he seems to have perceived a personal attack, as Buñuel himself commented that Lorca felt alluded to as the target of the title, thinking he was the "Andalusian dog."

What *Trip to the Moon* and *An Andalusian Dog* share above all else is the attempt to incorporate poetic discourse into the cinema, in other words, to transform the poetic image into visual image, merging in this way two different conceptions of poetry: as a mode of literature and as a property that can be shared by the different arts.

The new art of the cinema, with its strong realistic legacy, reveals the separation between the poetic word and visuality as such. Lorca's strategy will be to apply to the visual image nonrational criteria of association, of unexpected conjunction of distant and alien elements, similar to the kind he practices in his writing. The "hechos poético-cinematográficos" with which he constructs *Trip to the Moon* are juxtapositions of unconnected images, which break with any continuity of space and time. They are images that can only be read with the resources of poetry, because their only law is Lorca's poetic logic.

Only very recently have we been able to enjoy a cinematic version of *Trip to the Moon*, directed by the artist Frederic Amat.[15] For years our desire to see had to be partially satisfied by looking at those other manifestations of Lorca's visual imagination, his drawings. In fact, Lorca anticipated the use of some of those drawings in the film, such as the already quoted "drawing of a head vomiting and opening and closing its eyes," or the direct reference to "a drawing: *Death of Saint Rodegunda.*" This title belongs precisely to a drawing by Lorca dated in New York in 1929, at the time when the screenplay was written, that also connects with the "head vomiting and opening and closing its eyes."[16]

The drawings, together with the many echoes of other works of the New York period we find in *Trip to the Moon*, illustrate a productive dialogue between the arts, and the artists, that necessarily determines our interpretation of this screenplay. To make a film Lorca resorts to drawing and, especially, to poetry, and takes advantage of the images and motifs he shares with his friends and of others that originate in his own poetic universe.

Lorca appropriates the resources and conventions of film, and he adapts them to his own themes and concerns, and to his poetics. He shapes his repertoire of poetic images to make them visible, to the point of turning to his own previous experience in the visual arts. This adaptation of the cinematic medium to his poetry, and of his poetry to the demands and limitations of the new medium, results in a unique product, close in some aspects to other works of the New York period but differing from them because of its double allegiance: to the visual image and the poetic image.

The rules of the game change with each image; they also change with each art. But there is a common rule, shared by the three artists during this period and described by Dalí in his 1928 article "Reality and Super-reality":

> Those things that are more alien in their potential for loving each other are linked in an interweaving of perfect love. Those things that are impossible to confuse are casually interchanged. (*AD*, 130)

We could call it the rule of poetry. It is that surplus of meaning left after all explanations are exhausted. It is the familiar object made unfamiliar by an extraordinary association. It allows us to confirm that when Lorca, Buñuel, and Dalí were dropping the word *poetry* all over the place, they were not being careless. They were actually being strictly faithful to the etymological roots of the word, since in classical Greek *poiesis* meant "fabrication," "production." Cutting and pasting with the signs at hand in any given art, even when what you cut happens to be a hand or an eye.

Notes

Unless otherwise indicated all translations from the Spanish are my own.

1. Agustín Sánchez Vidal, *Buñuel, Lorca, Dalí: El enigma sin fin* (Barcelona: Planeta, 1988); Ian Gibson, *Federico García Lorca* (Barcelona: Grijalbo, 1985); Mario Hernández, "García Lorca y Salvador Dalí: Del ruiseñor lírico a los burros podridos (Poética y epistolario)," in Laura Dolfi, ed., *L'imposible/posible di Federico García Lorca* (Naples: Edizione Scientifiche Italiane, 1989), 267–319; Rafael Santos Torroella, *La miel es más dulce que la sangre: Las etapas lorquiana y freudiana de Salvador Dalí* (Barcelona: Seix Barral, 1984); and *Dalí residente* (Madrid: Residencia de Estudiantes, 1993).

2. Federico García Lorca, *Viaje a la luna*, ed. and intro. Antonio Monegal (Valencia: Pre-Textos, 1994). Hereafter *VL*, cited in the text.

3. Luis Buñuel, *Obra literaria*, ed. Agustín Sánchez Vidal (Saragossa: Heraldo de Aragón, 1982), 145–48.

4. Mario Hernández, ed., *El libro de los dibujos de Federico García Lorca* (Madrid: Tabapress, 1990), translated as *Line of Light and Shadow: The Drawings of Federico García Lorca*, trans. Christopher Maurer (Durham, N.C.: Duke University Press, 1991).

5. Federico García Lorca, *Epistolario*, ed. Christopher Maurer (Madrid: Alianza, 1983), 2:74.

6. Jean Laude, "On the Analysis of Poems and Paintings," *New Literary History* 3, no. 3 (1972): 476.

7. Salvador Dalí, *L'alliberament dels dits: Obra catalana completa*, ed. Fèlix Fanés (Barcelona: Quaderns Crema, 1995), 41. Hereafter *AD*, cited in the text.

8. Buñuel, *Obra literaria*, 184–86.

9. Hernández, ed., *Line of Light and Shadow*, 148.

10. Federico García Lorca, *Obras completas*, ed. Arturo del Hoyo (Madrid: Aguilar, 1989), 3:160. Hereafter *OC*, cited in the text.

11. Michael Riffaterre, "Semantic Incompatibilities in Automatic Writing," in *Text Production*, trans. Terese Lyons (New York: Columbia University Press, 1983), 221–39.

12. Federico García Lorca, *Conferencias*, ed. Christopher Maurer (Madrid: Alianza, 1984), 1:101.

13. Ibid., 2:17.

14. Buñuel, *Obra literaria*, 141.

15. Visual documentation of Amat's drawings for the storyboard, stills from the film, and pictures of the shooting, in addition to the text, can be found in Federico García Lorca-Frederic Amat, *Viaje a la luna* (Barcelona: Edicions de l'Eixample-Centre de Cultura Contemporània de Barcelona-Comisión del Centenario de Federico García Lorca-Institut d'Edicions de la Diputació de Barcelona, 1998).

16. The drawing has been catalogued as number 161 by Mario Hernández in *Line of Light and Shadow*, which includes, as number 162, another version of the same subject.

The Image in a Fatal Kiss: Dalí, Lacan, and the Paranoiac Representation

Pithamber R. Polsani
University of Arizona at Tucson

The perverse image seemed both and neither.

—Dante, *Inferno*

GOLDEN luminosity consumes the barren plain and the cloudless sky. The horizon barely visible in the yellow luminescence appears to be in flight. On this desolate terrain, in the corner, a solitary translucent amphora—broken—casts an incongruent shadow. In the play of splendid light and pale shadows, blurred vision of a distant town discloses only the contours of congregated buildings. A single two-wheeled cart with two seated figures approaches the settlement.

Bright landscape with a faraway town and a cart with two figures is the subject of this simple painting: *The Phantom Cart* (1933). In its simplicity the phantom cart captures the gaze of the viewer, demanding a second look. Now, what were wheels have become two staffs lodged in the ground; the two figures in the cart are silhouettes of two towers in the faraway town. The cart that was en route to its intended place has merged with the point of its arrival. Now, with the erased itinerary, the cart is an image of an incessant motion neither traversing space nor passing time. Without space to cross and time to conquer, the cart is its own destination.

The Phantom Cart is about transformed space-time wherein space is neither a point (Euclidean), a dimension (Cartesian), nor a pure intuition (Kantian) but "material and real" (Einsteinian). It is no longer a container of flesh but flesh that is at once the contained and the container. Space is an envelope of viscous consistency that is "colossally alluring, voracious, and personal, that presses at every moment with its disinterested and soft enthusiasm the smooth finesse of the 'strange bodies.' " This gelatinous space spouts new aerodynamic and atmospheric "Beings-Objects," much as blackheads are squeezed from the surface of a nose. These new beings-objects, "flattened, bumpy," and "with their super-smooth curves and . . . their massive salivary anatomy," resemble "pot-bellied thighs and sagging bellies."[1] The aerodynamism of the new space obliterates the difference between objects and beings, giving rise to a new period of softness: soft watches, soft automobiles, floating kidneys, and liquid engines. Hence the surrealist apartment shaped like Mae West's face: the golden locks of the actress's hair adorn the oval entrance of the apartment like curtains; her eyes are two paintings on the back wall; her nose, the fireplace; and her lips, the solitary sofa in an empty room carved like her cheeks. These beings-objects or objects-beings are captured and frozen in the moment of their metamorphosis. Thus they are perverse and polymorphous—like the legs of the chair and table in *Woman with a Head of Roses* (1935)—which are shaped like a woman's legs and are exceedingly long and slender, conforming to the contours of the base of the table and the chair that have fortuitously come together. If objects and beings share the identically named anatomical parts, then anatomies themselves are interchangeable since both inhabit aerodynamic space.

The undulations of soft space produce images and their doubles that appear in the form of phantoms and specters. In the play between the immobile stability of a phantom and the luminous rapidity of a specter emerges a series of beautiful canvases. *The Phantom Cart,* one of those canvases, fuses voluminous phantom towers and slender spectral cartwheels into a single image that evokes two sensations: motion and inertia, neither one transforming into the other. That is to say, the mobile cart will never reach its destination, nor will the faraway buildings ever be approached. It is the scenario of perpetual mobility and insistent inertia.

> The imitator, being the creator of phantom,
> knows nothing of reality.
>
> —Plato, *Republic*

"It is by a distinctly paranoiac process," says Dalí, "that it has been possible to obtain a double image: in other words, a representation of an object that is also, without the slightest pictorial or anatomical modification, the representation of another entirely different object, this one being equally devoid of any deformation or abnormality disclosing some adjustment" (*CW,* 224). The paranoiac critical activity is not a passive state like hallucinations, dreams, experimental oneirism, and psychic automatism already in use in the surrealist experimentation; instead it is an active state that through its power and violence uses coincidence and pretexts to produce the second image. It is a "*spontaneous method of irrational knowledge based on the interpretative-critical association of delirious phenomenon*" (*CW*, 267). Dalí proposes this alternative method of pictorial representation because the surrealist experimentation with dreams, psychic automatism, and other phantasmic phenomena lends itself to psychoanalytic interpretation which in turn reduces it to rational and logical discourse, overlooking the "uninterpretable residue . . . and very vast margin of enigma" (*CW*, 266) that resides in such phantasmic images. In paranoiac-critical activity "the presence of the active and systematic elements peculiar to paranoia guarantees the evaluative and productive character" of the new delirious images belonging to the irrational, propelling them "toward their physical and real 'possibility': they go beyond their realm of psychoanalyzable phantasm and 'virtual' representations" (*CW*, 266). Since the delirium of paranoia, the foundation of paranoiac-critical activity, already possesses a systematic structure, any explanation of their representation, psychoanalytic or otherwise, is a posteriori phenomenon, a superimposition on a painting that already exists as a phenomenon. Therefore, Dalí declares, "the fact that I myself, at the moment of painting, do not understand the meaning of my paintings does not indicate that these paintings have no meaning: on the contrary, their meaning is so deep, complex, coherent, involuntary, that it escapes the simple analysis of logical intuition" (*CW*, 265).

The purpose of paranoiac-critical activity is to discredit the world of reality by meticulously sowing confusion between the rational and the irrational, the real and the delirious. This subversive act is achieved by depicting the images of concrete irrationality with a realism of fastidious precision, approximating the illusionism of trompe l'œil and "with the corresponding means of expression approaching those of the great realist painting—Velázquez and Vermeer of Delft—in order to paint realistically according to irrational thought, according to the unknown imagination" (*CW*, 265).

Dalí translates this technique masterfully to several canvases he painted in the thirties exploring to the maximum the irrational delirium of the paranoiac with multiple images. In one of the early paintings of this kind, *Invisible Sleeping Woman, Horse, Lion* (1930), the central figure is an envelope of three figures listed in the title, wherein the reclining woman's head of hair is also the horse's mane and her body transforms into the horse's legs. In this painting Dalí has not yet mastered the skill to render multiple images in a single configuration. The cinematic version of the same scenario, unrealized, consisting of six images, shows clearly how this technique in a mobile form could be successful. However, the subsequent canvases, such as *The Slave Market with a Disappearing Bust of Voltaire* (1940), exploit ingeniously the potentiality of multiple images of a delirious phenomenon. In this painting, Gala, seated on the left, gazes at the archway of ruins wherein various women standing in the archway converge, generating a double image that oscillates between the bust of Voltaire and the figures. Voltaire's face is made of two black women dressed in black and white in the fashion of seventeenth-century Spain: their faces become his eyes, their lace collars form his cheekbones, their gloved hands his jaws, and the gaping aperture in the archway his head. In *Spain* (1938) the face and the chest of a woman gazing melancholically into a void is a congregation of figures in battle, her red lips being formed by the red clock of a standing figure and a figure on horseback. The bodies of two soldiers on horseback in battle, one piercing his enemy with a lance, and the other mortally wounded, falling to the ground, form the contours of a woman's breasts, and their draped heads form her nipples. These battle scenes that make up the face of the standing woman in the painting are inspired by Leonardo da Vinci's sketches of cavalry combat. Moreover, Dalí sees the paranoiac-critical method already existing in the Italian master's work: "Leonardo proved an authentic innovator of paranoiac painting by recommending to his pupils that, for inspiration, in a certain frame of mind, they regard the indefinite shapes of spots of dampness and the cracks on the wall, that they might see immediately rise into view, out of the confused and the amorphous, the precise contours of the visceral tumult of an imaginary equestrian battle."[2] The painting *Spain*, rendered in the middle of civil war, is perhaps a symbolic evocation of her mutilated history. Similarly, in the *Great Paranoiac* (1936) two faces, the large one in the foreground and a small face in the background, are a collection of faceless, tortured figures, their contorted bodies evoking at once the postures of dancers in a choreographed mise-en-scène as well as a group of patients persecuted by their delirium in a medieval asylum. One of the most elaborate canvases that depicts on a large scale the multiple images is *The Endless Enigma* of 1939 in which six large images—mythological beast; face of the great one-eyed moron; reclining philosopher; greyhound; mandolin, fruit dish with pears and two figs on the table; seated woman mending sails—are contained in a single composition. When this painting was exhibited in the Julian Levy Gallery in New York (1939), the exhibition catalogue carried six drawings that broke the composition into its constitutive parts.

The paranoiac-critical activity that generates multiple images is not "concocting entertainments on the level of crossword puzzles," as André Breton wrote in "The Most Recent Tendencies in Surrealist Painting"; instead, the force of paranoiac-critical activity resides in the central obsessive image which is gratuitously generated by the mind and around which other images suggested by the same paranoiac delirium are grouped. The obsessive image itself is, as Dalí aptly characterizes in *The Rotting Donkey*, a simulacrum. And as a simulacrum, the image, although it belongs to reality, has been severed of all its relationships with its environs. The simulacrum is an image emptied of all the representational value normally attributed to an image. Therefore, "the violence and the traumatic essence of the simulacra with regard to reality, and the absence of the slightest osmosis between reality and the simulacra, . . . lead us to infer the (poetic) impossibility of any kind of *comparison*" (*CW*, 225). The paranoiac image as simulacrum, independent of content, reference, and value can overrun reality by assuming its form, and reality in turn tries to imitate the simulacra.[3] Furthermore, the images of paranoiac-critical activity cannot be subjected to a rational analysis nor can they be translated to a logical language of psychoanalysis because they constitute an active system. It is in the characterization of paranoiac image as simulacrum and the delirium as a system that Dalí's method of paranoiac-critical activity has its closest affinities with Jacque Lacan's analysis of paranoia.

The paranoiac observes strange happenings in the street, and either they are favorable to the subject or unfavorable, but they cannot be comprehended or logically expressed. For example, a red car parked on the street is not a chance encounter but something that has been deliberately placed there for some purpose. Although the happenings are mysterious and cannot be articulated, the paranoiac insists that the events have meaning, and within the subject's delusion this meaning is given to understanding; perhaps it is even given to understanding of the subject's interlocutors, but what is "striking is that it [meaning] is inaccessible, inert, and stagnant with respect to any dialectic." The peculiar character of this meaning is its repetitiveness, its insistent reiteration with "one interrogative sign"[4] without its integration into a dialogue or an answer. The meaning that appears as overwhelming in the delusional discourse that Lacan refers to is not signification, which is an operation of discourse on the diachronic order of signs, but the signified, in the dyad of signifier/signified that forms the sign. The meaning (signified) is not an indication of objects and beings of the material world but a mental image, as Ferdinand Saussure calls it in the *Course in General Linguis-*

tics. Therefore, the function of meaning (signified) is to refer to another meaning, while its value is derived from its negative relation to another which it is not.

The paranoiac subject introduces disorders and alterations in the synchronic order of the signifier through which some signifiers undergo a modification, acquiring a value, "a particular force of inertia, become charged with meaning" (*P,* 54). In delusional discourse the signifier[5] acquires two peculiar characteristics. First, as a neologism the signifier of the paranoia provokes an exceptional discordance in ordinary language; and second, the signifier acquires meaning, not the meaning that refers to another meaning, but a meaning as such: "It's a meaning that essentially refers to nothing but itself, that remains irreducible. The patient himself emphasizes that the word carries weight within itself. Before being reducible to another meaning it signifies within itself something ineffable, it's a meaning that refers above all to meaning as such" (*P,* 33). This meaning that refers to nothing is repeated and reiterated with stereotypical insistence. For example, Daniel Paul Schreber in his memoirs, *The Memories of My Nervous Illness,* insists on the overwhelming importance certain words, such as *Nervenanhang, nerve-contact* and *Seelenmord, soul murder* have, and these words "have been repeated to him in their elective meaning which he does not always understand terribly well. *Seelenmord, soul murder,* for example, . . . is problematic for him, but which he knows has a particular sense" (*P,* 54). But, Schreber talks about all this in an ordinary discourse, communicating it in a very lucid manner to his readers. Herein lies the peculiar nature of delusional discourse; it is an ordinary landscape overlaid with insistent signifiers that assert again and again their meaning. Dalí masterfully renders these landscapes on canvases.

In a series of images painted in the 1930s, *Paranoiac Astral Image* (1934), *Apparition of My Cousin Carolineta on the Beach at Rosas* (1933, 1934), *White Calm* (1936), *Mediumistic-Paranoiac Image* (1935), and *Sun Table* (1936) ghostlike figures appear on a prepared background. These poetic paintings, fortuitous encounters, could only have been compiled by a paranoiac activity that dislocates images of reality from their habitat and reprojects them with intensity on a real plane. The neologism and the ineffable meaning Lacan analyzes as the characteristics related to paranoiac delusion can be clearly perceived in the brilliant figures dressed in nineteenth-century fashion that emerge on these canvases. Dalí called these figures "instantaneous": hallucinatory sequences with stereoscopic images projected on a "real" setting wherein the figures in themselves, between them, or in relation to the backdrop do not offer any narrative that can be called the "meaning."

While the signification is extraneous to paranoiac scenes, there is a system and systematization inherent to it. In any system the elements that constitute it are related through their differences and similarities with each other. The value of a sign in a linguistic system, for example, is "entirely relative" and it resides in each sign's relation with others' "simultaneous presence." In a system of signs the values are governed by a paradoxical principle of dissimilarity and similarity. A "dissimilar thing . . . can be exchanged for the thing of which value is to be determined" and a "similar thing . . . can be compared with the thing of which the value is to be determined."[6] In Dalí's paintings disparate pictorial elements are brought together into relation as phantoms with volume and as specters devoid of volume.

The extremely thin female figure in two canvases painted in 1933 and 1934 with the same title, *Apparition of My Cousin Carolineta on the Beach at Rosas,* floats on the beach gazing at a lonely cypress and heavy clouds in the sky. While the heavy clouds impending on the whole scene are voluminous phantoms, the female figure and the lonely cypress reduced to their bare features represent the specters. This relation constituted through volume and absence of it generates a system; nevertheless, the very idea of a system itself is questioned by revealing the simulacral and virtual nature of volume: "The simulacrum of volume is the envelope.—The envelope hides, protects, transfigures, stirs up, tempts, gives a misleading notion of volume.—It causes ambivalence with regard to volume and makes the volume suspect" (*CW,* 203). Dalí's paintings are founded on two contradictory principles that render paintings totally irrational, devoid of meaning. On the one hand, a relationship is established between figures through oppositions and similitudes; on the other hand, the very idea of contrast and resemblance is made suspect.

In *White Calm* the human figures are meticulously drawn in luminous tones approximating a real scene on a beach, exactly copying a postcard Dalí found in Cadaqués. This painting brings to practice what Dalí declared as his objective in art in *The Conquest of the Irrational*: "My whole ambition in the pictorial domain is to materialize the images of concrete irrationality with the most imperialistic furor of precision, so that the world of imagination and concrete irrationality may be of the same objective clearness . . . as that of the external world of phenomenal reality" (*CW,* 265). However, by introducing a lone amphora, which lies in the extreme bottom corner of the canvas, at the threshold where the painting turns into a postcard, the painter disturbs the realism of reality.

The reappearance of a broken amphora in several pictures of this period is similar to the double image. The insistent repetition of the same image in multiple scenarios, a fundamental feature of paranoiac delirium and paranoiac-critical activity, is not extraneous to the subject but has its origin in the formation of the ego through the perception of an image in a mirror.

> I've so dreamed of you
> That you've lost your reality
> I've so dreamed of you
> so walked, spoken, slept with your ghost that nothing remains of me perhaps,
> and yet, but to be a ghost among ghosts and a hundred times more
> like a shadow than the shadow that strolls and will stroll lightly on the
> sundial of your life
> —Robert Desnos

The child is jubilant at discovering his image in a mirror. Not yet capable of standing or moving on his own, still reeling under the prematurity of his birth, the child is enraptured by the mirror image that reflects to him his body as unity, which he has experienced so far only as fragmented and uncoordinated drives and sensations. This initial relation to imago is formative of an Ideal-I, the "symbolic matrix" that would lead to its consolidation in the dialectic of identification with others and the subsequent formation of the child as subject by language. The euphoric assumption of an image is not a discovery in reflection of an already existing body, but the reverse is true. That is, through representation in the mirror the child comes to recognize himself as an individual body like that of others he sees, and he therefore comes to know that his is also a body among others. This body is not yet given, but is a promise for the future, an anticipation of a fully formed and unified body that will overcome the present immobility and incapacity. Moreover, "the total form of body by which the subject anticipates in a mirage the maturation of his power is given to him only as *Gestalt*, that is to say, in an exteriority in which this form is certainly more constituent than constituted."[7]

Through identification with the mirror image and the image of the counterpart that anticipates his mastery of motor control and freedom, the child is outside of himself. He is constituted in a visual space by the gaze which is the threshold of the visible world. This formation of ego and the image of body dependent on the image of the Other, the double, gives rise to two correlative tendencies which are the essence of ego: narcissistic love and paranoiac aggression. The Other's image attracts and repels at the same time. The one I love is also my rival whom I hate. The mirror stage is a mise-en-scène which gives rise in the subject—who is caught in the "lure of spatial identification"—to the fantasies of a fragmented body as well as to a form of its totality—the orthopedic body. The phantasm of fragmented body appears in dreams "in the form of disjointed limbs, or of those organs represented in exoscopy, growing wings and taking up arms for intestinal persecutions—the very same that the visionary Hieronymus Bosch has fixed, for all time, in painting, in their ascent from the fifteenth century to the imaginary zenith of modern man."[8] The only solution to this discord between love and hate, the unending oscillation between unity and dismemberment, is the passage to action: the paranoiac suicidal gesture of killing the aggressor. Therefore, Aimée attacks her favorite actress, Madame Z, at the entrance to the theater and the Papin sisters cut to pieces their mistresses' bodies.[9] It is suicidal because in killing the Other, the paranoiac kills himself or herself. In the deathly blow struck at the Other, the image is shattered to pieces, thus returning the subject to a pre-mirror stage. Perhaps we can say, contrary to conventional interpretation, Narcissus takes the fatal step and plunges to his death not out of love for his own image but out of hatred. The revulsion of the imago in paranoiac acts signals the failure of the mirror-image phenomenon, that is the identification with the image as the foundation of narcissism.

The pictorial representations of dreams of a dismembered body, the paranoiac suicidal act, and the constitutive gesture and predominance of imago can be found in Dalí's canvases. The multiple images generated by the paranoiac-critical activity are akin to the fantasies of the fragmented body and its correlative, the unified orthopedic body. For example images in the paintings *Spain* and *Great Paranoiac* portray multiplicity as well as unity, appearing at once as shattered fragments and as totality. The multitude of voices the paranoiac communicates with and the innumerable characters that persecute and taunt the paranoiac are none other than fragments of a single image, that is, the ego of the subject. In Schreber's delirium the souls invade his body; the host of characters he calls "fleeting-improvised-men" (*P*, 97) appear before him, and he considers himself only as a second copy of his own identity. These "others" Schreber refers to as "they" are none other than his ego as the Other that "is structurally multipliable and divisible" (*P*, 98). The subject is the creator of this Other in fragmented form and he is attached to this Other "not as object, image, or shadow of an object, but to the other in his essential dimension . . . to this other who is irreducible to anything other than the notion of another subject, that is, the other qua he." In delirium, as Lacan observes, "the subject speaks *to himself with* his ego" (*P,* 101, 14).

Dalí portrays the suicidal gesture of paranoiac aggression in two disturbing paintings of 1936: *Soft Constructions with Boiled Beans-Premonition of Civil War* and *Autumn Cannibalism.* In the first one, a grotesque figure on a barren rocky plain tears apart its own body with violence, and in the second painting two human figures using a spoon, knife, and fork devour each other's flesh. Referring to this painting, Dalí observed: "these Iberian beings mutually devouring each other correspond to the pathos of civil war considered as a pure phenomenon of natural history as opposed to Picasso who considered it a political phenomenon."[10]

But, the subject of most consistent and imaginative exploration in Dalí, visually as well as theoretically, is the fragmented body. It acquires a very revealing and prophetic dimension in his essay "The New Colors of Spectral Sex-Appeal." Continuing with the argument of tenuous opposition between phantom and specter, he takes a curious turn to explore the future of woman's sex appeal. The new sexual attraction of woman will consist in the exploitation of her spectral qualities and resources, "their potential carnal and luminous dissociation and disintegration." The new spectral woman "will be the *dismountable woman*" (emphasis added). Her anatomy subjected to disfiguration and disarticulation will render her spectral, and her dismountable body will become the subject of most highly refined analytical methods, which will allow pieces of her anatomy to be exhibited separately. This dismembering, inquisitive procedure will be accomplished through "aerodynamic costumes and irrational gymnastics." New and artificial anatomical parts will be invented to accentuate "the atmospheric sensation of breast, buttock or heel, . . ." and the new false breasts "will be extremely soft and nicely molded. . . ." Further, the new dismountable woman will have a spectral smile that "will be artificially prompted by vibratory metallic fibers in hats" (*CW,* 206–7).

The incessant images of sexuality that press upon us today, the ineffaceable reality of surgically remaking anatomies, obsessive devotion to exercise, individual machines tailored to reshape each anatomical part . . . perhaps we should let Dalí's words resonate with their oracular power.

Reality is a bitch.
—Jean Baudrillard, *The Perfect Crime*

The double image either as an envelope of multiplicity or as an image shattered into shadows; the recurring image that returns with insistence in the most unexpected space and at an unintended hour; the embracing phantoms and specters that lay bare the false opposition between the two; the infinite morphing of being-objects, all are related in their fixation, in their being as obsessive images, in their repetition and plasticity. What is the essence of image that engenders these endless duplications and reduplications?

The obsessive image is a pure form and it is in form that the multiple images resemble each other, enact similitudes, abolish contrariety and obscure distinctions: the endless repetition of the same. In *Spain* and the *Great Paranoiac* the form of the face is a formal grouping of forms that are contained in it and overflow it, establishing at the same time the confines of this form (face) and yet revealing the pure formality of these thresholds. Narcissus seated, with head in repose on his bent knee in *The Metamorphosis of Narcissus* (1936)[11] is repeated in the two fingers holding a cracked egg from which blooms the narcissus flower. The self-referential images of painting are rehearsed in the grapheme:

> Narcissus, in his immobility, absorbed by his reflection with the digestive
> Slowness of carnivorous plants, becomes invisible.
> All that remains of him
> is the hallucinatory oval of whiteness of his head,
> his head more tender again,
> his head, chrysalis of biological after-thoughts,
> his head held up at the tips of the fingers of water . . .

The golden luminescence of Narcissus' *eidos* (the visible aspect) is an anticipation of its own repetition as a reflection. The imago and Narcissus coincide insofar as they duplicate each other in their visibility as forms. But Narcissus takes the fatal step.

The same gesture is reenacted in *Morphological Echo* (1934–36), *Nostalgic Echo* (1935), *Anthropomorphic Echo* (1937), *Landscape with Girl Skipping Rope* (1936), and *Suburbs of Paranoiac-Critical Town: Afternoon on the Outskirts of European History* (1936). The semblance of the girl with hoop, reminiscent of Giorgio de Chirico's running child in *Mystery and Melancholy of a Street* (1914), is repeated in the contour of the swaying bell in the tower embrasure. The hollow in the back of the seated Hitlerian nursemaid in *The Weaning of Furniture Nutrition* (1934) is equivalent to the form of the table in front of the maid.

The table in turn has a portion of it cut out in the shape of another small table placed next to it with a bottle on top. This duplication is taken to its finest manifestations in two paintings: *Morphological Echo* of 1936 and *Suburbs of Paranoiac-Critical Town: Afternoon on the Outskirts of European History*. In the first, on a table (as altar) in the foreground are three innocuous objects: a glass with a spoon, a crusty end of a loaf of bread, and a bunch of black grapes. The background, monochromatic with subtle gradation of color, is devoid of any point of reference; however, the perspective is achieved with the manipulation of sizes of objects and figures. These objects on the table lead vertically to corresponding forms and those in turn to three other objects in the distance. The glass with a spoon on the left leads the eye to a walking figure and to a tower; the bunch of grapes on the right find their analogous form in the reclining figure of Lenin, and the figure in turn is complementary to a wall. The bread in the middle of the table is echoed in the figure of a seated nurse, who is related formally to a rock in the distance.

The *Suburbs of Paranoiac-Critical Town: Afternoon on the Outskirts of European History* renders much more complexly interlocked images that are formally related to each other. The canvas is divided into four distinct spaces of which three are architectural and the fourth is a grouping of objects and figures in a random order. Each space is related to others through the contiguity of forms that repeat and reiterate themselves. Moreover, forms, figures, and objects from other paintings which reappear not only forge the link on a formal level between *Suburbs of Paranoiac-Critical Town* and other works but also disclose the inner nature of the delirious phenomenon. Just as a paranoiac carries his signs everywhere, the painter moves from canvas to canvas with the same figures repeating again and again as if carrying out the same painting again and again.

On the extreme right of *Suburbs of Paranoiac-Critical Town* a doorway with a semicircular arch leads to a street with two human figures, one seated in front of a door and the other bent over, picking up something from the street. At the end of the street there is another doorway, circular in shape, leading to another space. In the center a triangular archway leads to a village with a prominent bell tower in the background. The girl with skipping rope from several earlier drawings and paintings reappears and is related to the swaying bell in the tower embrasure. The form of a ball on top of the tower is linked to the circular opening on top of the triangular entranceway and the doorway in the background of a street scene. Further, the semicircular archway under the tower leading into the village is identical to the one on the right leading to a street. The third architectural space is a reproduction of a building in Palamos and whose dome is related to the mirror on the dressing table in the center of the canvas.

In the central section of the painting the form of a bunch of grapes in Gala's hand is linked to a skull on the table and the hindquarters of the horse behind (the work drawings for this painting show various ways in which Dalí tried to forge the links). The shape of the keyhole of a small table, reminiscent of the keyhole in *The Weaning of Furniture Nutrition,* is identical to the key that appears at the top, while the imprint on an empty chair echoes the draped figures standing behind it. These figures in turn forge a connection with phantasmal impressions of absent bodies in an earlier painting, *Surrealist Composition with Invisible Figures* (1926 and 1934–36).

As is evident, the paranoiac-critical activity finds its finest and most complex rendition in the *Suburbs of Paranoiac-Critical Town: Afternoon on the Outskirts of European History.* By linking various objects, figures, architectural elements, and impressions through the form of their appearance, Dalí divulges the inner essence of paranoiac delirium and its creative force. That is, the obsessive image that reiterates and repeats with insistence in paranoia and the affective investment of a paranoiac in his obsessions is not associated to any content of the image but to a form, and as a form alone the image is given to its reiteration. The obsessive image of a paranoiac is rational on the one hand because of the consistency in its reappearance, and yet it is irrational on the other hand because of its total disregard for the materiality of an object or figure on which it leaves its mark. In the form, the rational and irrational converge, disclosing the tenacity of their opposition. Furthermore, the imago of mirror stage—the foundation for the formation of human ego as the Other—that resurfaces as obsessive image in the delirious phenomenon was always already a form. As a pure reflection on an unruffled surface it is devoid of any content. Lacan insists on this formal nature of imago: "we the psychoanalysts are here reintroducing an idea discarded by experimental sciences, i.e., Aristotle's idea of *Morphe.* In the sphere of relationships insofar as it concerns the history of the individual, we only apprehended the externalized images."[12]

Salvador Dalí's paintings of the 1930s explore with great subtlety various facets of paranoiac-critical activity whether it be the hallucinatory nature of painted surface, double image, or obsessive image, or the continuity of a form in several objects and figures. Though the technique of painting Dalí employs can be characterized as visionary automatism wherein the painter faithfully transcribes images to a canvas that he has dreamt or hallucinated, the ultimate objective is not a translation of images earlier experienced but the employment of this technique and utilization of the creative force of paranoiac delirium to problemitize the notion of representation he had inherited in the Western tradition. Dalí was not alone in this endeavor; all his contemporaries—Pablo Picasso, André Breton, Giorgio de Chirico, Max Ernst, Paul Eluard, Marcel Duchamp, Guillaume Appollinaire, and others—conceived their projects as a movement against representation: a representational art that accords priority to signification.

Representation has occupied the most prominent place in the conceptualization of the world (world being that which is only given in representation) and knowledge since the seventeenth century. Within the general domain of representation, the sign whose function is to represent is the fundamental element of analysis and the constitution of knowledge of the world given in representation. A sign consists of two entities: a signifying element (the thing that represents), and a signified element (the thing represented). Although both elements have the same status as representations, in analysis the signified, misconstrued as "meaning," takes precedence over the signifying element. Therefore, a sign is charged with the responsibility of representing what is considered "reality" either as an idea or an event. In the pictorial domain, a drawing or a painting is judged in terms of its accuracy in rendering the idea behind it and adequateness to the reality that sustains it. Hence the painting becomes symbolic of some other object or idea.

Saussure, in his project for a general semiology that rescues language from being reduced to a simple naming process, brings into sharper focus the relationship between the signifying element and the signified by arguing that their encounter in the constitution of a sign is purely arbitrary. There is no inherent relationship between a signifier and its signified that necessitates their coming together. Saussure's definition of the signifier as "sound-image" and the signified as "concept" is akin to the classical definition of the signifying element and the signified element.

Saussure represents graphically the relationship between two elements of a sign, in a fatal gesture, by placing the signified over the signifier:

Concept
Sound-image

This has led to the misunderstanding that the signified has preference over the signifier and that the concept understood as meaning takes precedence over its phonic representation. Therefore, signification in discourse is the ordered signifieds represented by the order of signifiers. However, signification is not inherent to the sign as unity of signified and the signifier but belongs to the diachrony of signs ordered by discourse. The confusion with regard to signified as meaning is already given, first, in conceptualizing sign as dyad, and, second, in representing the relationship between the two as signified over signifier. Furthermore, the confusion has its roots anterior to the division of sign into two elements, that is, in conceiving signs either as representing reality, psychic or otherwise, or, conversely, signs as creators of reality. As long as signs stand in relation to reality, signification impinges upon them with violence.

Salvador Dalí, who stated his intention of sowing systematic confusion in reality, was aware of this double bind in signs; therefore, pictorial elements in his paintings following the paranoiac-critical method are signs extracted from their abode and liberated from the function of representing or creating reality. Although the signs in the paintings are related within a paranoiac system, the system itself is rendered suspect and the signs are liberated for the second time: from the very notion of relationship. Like Prometheus who stole fire from the gods and gave it to humans, Dalí robs reality of its signs and offers them to simulacrum. The rotting donkey evoked in his 1930 essay is none other than reality. In the absence of signs that would reflect it like an image in a mirror, reality is condemned to slow decay and agonizing death. Deprived of signs and signs turned to simulacrum, reality makes one last desperate attempt to imitate simulacra, but the simulacrum is indifferent to reality's plight.

Notes

1. Salvador Dalí, "Aerodynamic Apparitions of 'Being-Objects,' " in *The Collected Writings of Salvador Dalí*, ed. and trans. Haim Finkelstein (Cambridge: Cambridge University Press, 1998), 209. Hereafter *CW*, cited in the text.

2. Salvador Dalí, Exhibition Catalogue (New York: Julian Levy Gallery, 1941). Quoted in James Thrall Soby, *Salvador Dalí* (New York: Museum of Modern Art, 1946), 21.

3. Dalí's theory of simulacra and its relationship to reality is akin to French philosopher Jean Baudrillard's characterization of contemporary culture as simulacral. For intellectual affinities between Dalí and Baudrillard, see this author's forthcoming article.

4. Jacques Lacan, *The Seminar of Jacques Lacan. Book III: The Psychosis*, ed. Jacques-Alain Miller, trans. Russell Grigg (New York: Norton, 1993), 22. Hereafter *P*, cited in the text.

5. Lacan employs the term *signifier* to mean sign, although in Saussure there cannot be signifiers independent of signifieds. The sign is defined as a combination of signifier/signified in which the signified is placed over the signifier. Lacan reverses this relationship by granting the signifier primacy over the signified. Furthermore, Lacanian analysis articulates

the possibility of a separation between the signifier and the signified: for example, in the case of the master signifier, the-name-of-the-father, which halts the sliding of signifiers in the diachronic order. As I will argue below, the above-mentioned separation is engendered more effectively by Dalí and other avant-garde artists.

6. Ferdinand Saussure, *The Course in General Linguistics*, ed. Charles Bally and Albert Sechehaye, trans. Wade Baskin (New York: MacGraw-Hill, 1966), 113, 114, 115.

7. Jacques Lacan, *Écrits: A Selection*, trans. Alan Sheridan (New York: Norton, 1977), 2.

8. Ibid., 5.

9. On 10 April 1930 a thirty-eight-year-old woman inflicted a knife wound on Madam Z's hand, a beloved actress of Paris, upon her arrival at the theater. The woman was admitted to the St. Anne mental hospital. Lacan observed her for over a year and half and called her Aimée. This case was at the center of Lacan's doctoral thesis of 1932, "De la psychose paranoiaque dans ses rapports avec la personnalité." Christine and Léa Papin, two sisters, worked as servants in a bourgeoisie family in Le Mans. One day the sisters killed their mistress and her daughter and mutilated their bodies. Lacan published an article on this crime, "Le Crime des soeurs Papin," *Le Minotaure* 3–4 (1933).

10. Quoted in Robert Descharnes, *Salvador Dalí: The Work, The Man*, trans. Eleanor R. Morse (New York: Abrams, 1984), 223.

11. Dalí took this painting to his only meeting with Freud in London, July 1938.

12. Jacques Lacan, "Some Reflections on Ego," *International Journal of Psychoanalysis* 34 (1953): 13–14.

Federico García Lorca's Andalusia in Light of Oswald Spengler's Theory of Magian Cultures

Alice J. Poust
Bucknell University

THE critical and popular success of Federico García Lorca's *Gypsy Ballads* in 1928 brought national attention to the young writer from Granada but also served to identify him, for part of the reading public, with certain stereotypical images about Andalusian culture which Lorca thought demeaning to his work. Always proud of his Andalusian heritage and sympathetic to Romany communities in Andalusia, Lorca ironically found himself in the position of having to clarify that he was not of Romany descent, as some readers assumed, and that those who thought the work was about Romany peoples or a celebration of the stereotypical, picturesque Andalusia had missed the point.[1] What, then, did Lorca have in mind? If much of the reaction to *Gypsy Ballads,* even the generally positive critical and popular reaction, was based on a degree of misunderstanding of his artistic objectives, how might we better understand those objectives? To what extent might the poet's portrayal of Andalusia aim at recasting the cultural significance of the region in light of contemporary theories of culture, rather than perpetuating the image of Andalusia as out of step with modernity, as such friends as Salvador Dalí and Luis Buñuel seemed to think? I shall argue that Lorca's portrayal of Andalusia in *Gypsy Ballads, Deep Song,* and *Divan of the Tamarit* tends to align Andalusia with a contemporary tendency to question and even oppose the construct "Western civilization." In doing so, I shall point to certain key similarities between the cultural ideas manifested in Lorca's writing, as poet and as critic, and the cultural theory formulated by Oswald Spengler in *The Decline of the West,* to demonstrate that Lorca's work effectively and systematically reinterprets Andalusia, not as a backward hinterland of Western or "Faustian" culture, to use Spengler's term, but as a part of an essentially different, "Magian" culture.[2]

The literary and artistic climate of the period between World War I and the Spanish Civil War was characterized by change and innovation, with one constant being the spirit of struggle between the established and the new. The poetry of Federico García Lorca, like the writing of most of his contemporaries, exhibits the same tendency to seek ever new forms of poetic expression. Numerous critics have recognized Lorca's use of avant-garde techniques in his writing, as well as the poet's generally positive sense of his Spanish and particularly Andalusian cultural tradition. While Lorca was often iconoclastic (we recall his portrayal of the Civil Guard in *Gypsy Ballads*, his satirizing of the middle and upper classes of his native Granada in *Gallo* [Rooster], and his criticism of the institution of the catholic church in *Poet in New York*), his poetry reveals a positive view of what we might call "traditional" Andalusian culture. This attitude toward Andalusian culture is not attributable, I believe, to an anomalous traditionalism in an otherwise progressive thinker. Rather, I would suggest, it is due to Lorca's sense that the cultural tradition to which Andalusia belongs might provide a counterpoint and, indeed, an alternative to the "Western" civilization whose values had come into question in the nineteenth century and particularly as a result of the destruction wrought by World War I. In this sense, Lorca's views of culture are consonant with his aesthetics.[3]

In *Decline of the West* Spengler treats cultures as complex organisms that are born in a particular geographical location, grow and develop, possibly come to maturity, and enjoy a period of plenitude, after which they enter into decline. Each culture is characterized by a particular "soul." The West, including primarily northern and northwestern Europe and English-speaking North America, is identified with the "Faustian" soul, a term Spengler borrows from Goethe. He associates the Faustian with the modern era, and particularly with the development and expansion of protestantism, technology, and capitalism, contrasting the Faustian soul with the earlier classical soul he attributes to the ancient Greek and Roman cultures of the Mediterranean. Between the flourishing of the classical and the Faustian cultures, he identifies a broad cultural tradition described as "Magian," linked to the early Christian church, particularly the Byzantine, and also to Judaism and Islam, although he traces the roots of Magian culture to earlier peoples in the Middle East, North Africa, and India, hence my view of Spengler's Magian cultures as "orientalist." It is possible that Lorca had some direct knowledge of Spengler's theory of cultural history, since the work was receiving attention in Spanish intellectual circles by the early 1920s, according to José Ortega y Gasset.[4] It also is possible that

Lorca and Spengler drew from some common sources. Ortega's preface to the Spanish translation of *Decline of the West,* in which he states, "it is a book that was born of profound intellectual needs and formulates thoughts that were beating in the breast of our time" (*DO* 1: 15), implies that Spengler's thought had been shaped by its historical and cultural context. Ortega even chides Spengler gently for taking credit for ideas that had been expressed previously by others (*DO* 1:15). While Ortega likely was referring to critical thinking in the realms of history and philosophy, it is clear that there were contemporary works in the realm of music, in particular, that expressed something akin to Spengler's description of Magian cultural values. Lorca's mentor Manuel de Falla had composed music evoking an Andalusia of fantasy and mystery in *Noche en los jardines de España* (Night in the gardens of Spain), featuring a stylized landscape, and also in *El amor brujo,* which opens with a scene in a gypsy cave near the sea. Both works were completed and would have been known to Lorca before the publication of *Decline of the West.*[5] The orientalism of Falla, of course, was not unique, as both contemporary and the previous generation of Spanish and other European composers had drawn from the musical traditions to be found in Islamic, Jewish, and Roma traditions. Lorca, himself, was writing poetry that contained references to Romany peoples, to Andalusian music and art, and featuring landscape imagery and cave dwellings as symbols before the publication of the Spanish translation of *Decline of the West* in 1923. While original, this work was not created in an artistic vacuum. The poetry of another Andalusian, Juan Ramón Jiménez, and the music of Falla, both of which often featured Andalusia, provided an aesthetic context for Lorca's *Deep Song.* Even so, by the time Lorca composed *Gypsy Ballads* (1924–1927), he had had the opportunity to become familiar with the cultural paradigms formulated by Spengler.

Although Spengler includes Islamic and Sephardic Spain within the Magian sphere, he does not deal with the Magian presence in Spanish or Andalusian culture in depth. Interestingly, we can find a model of how Spengler's thought might be applied to Spanish culture in the writings of a North American, Waldo Frank. Frank's work *Virgin Spain,* parts of which were published in Spain in the mid-1920s, provides insight into the contemporary cultural debates regarding Spain's relationship to Europe and to Western Civilization.[6] Even though Frank does not acknowledge a debt to Spengler, he almost certainly had read Spengler's *Decline of the West.*[7] Like Spengler, he recognizes the roles of classical and Germanic cultures in Spain and Andalusia's historical traditions but emphasizes the contribution of Jews, Muslims, and the more recently arrived Romany groups to Andalusian culture.[8] Spengler focuses on the music of the Faustian culture, especially from the Baroque to the nineteenth century, as he views this art form as reflecting the height of Western civilization (*DW,* 121). In keeping with his belief that the fullness of Magian culture is manifested in the ornamental arabesque, rather than in its music, he does not study the music of Magian peoples (*DW,* 109). Frank's attention to Andalusian music fills that void in Spengler's study. The same is true of Frank's recognition of the excellence of contemporary Spanish poetry and painting, as exemplified in the works of two Andalusians, Juan Ramón Jiménez and Pablo Picasso, respectively. Although Frank does not expressly use Spengler's term "Magian," his portrayal of Spanish poetry, art, and music tends to highlight the differences between the thought and the art forms of Spain from those associated with "the West," while emphasizing their "eastern" origins. In this sense, his interpretation of Andalusian culture as, in effect, Magian, may provide a bridge between the thought of Spengler and that of Lorca.

In this study I shall address some key similarities between Spengler's description of Magian culture and the creative and critical works of Lorca, particularly as they pertain to ethnicity, to symbolic representations of the world, to the idea of cultural essence, and to the relationship between culture and science. In doing so, I shall also point to the mediating presence of Frank's interpretation of Andalusian culture.

With regard to ethnicity, Spengler includes a wide variety of peoples within Magian culture, which could include ancient Babylon, Persia, Palestine, the classical Mediterranean cultures, the eastern part of the Roman empire, as well as Egypt and India (*DW,* 238). Magian culture is characterized by religious diversity, although Spengler highlights the early Christian, the Jewish, and the Muslim components (*DW,* 279–80). He particularly associates the flourishing of Jewish culture with the land that would become Spain (*DW,* 348–49) and considers Tarragona, Toledo, and Granada to be "predominantly Jewish cities" (*DW,* 349). He also recognizes the importance of Islamic hegemony in medieval Spain and Andalusia, as well as the Islamic legacy in Spanish culture. Turning to Lorca, it is clear that his portrayal of Romany people in *Gypsy Ballads* emphasizes the oriental component of Andalusian ethnicity in contemporary culture, while the aesthetic sensibility of *Divan of the Tamarit* evokes the Islamic legacy of Granada. As critic, Lorca also recognizes the importance of Jewish culture in Andalusia and, personally, liked to speculate that his second surname, Lorca, suggested a certain Jewish heritage of his own, according to Ian Gibson.[9] Lorca's recognition of the contribution of Jews and Muslims as well as of Romany groups to Andalusian culture is consonant with Spengler's description of Magian peoples.[10] I would suggest that for Lorca, Spengler's identification of Magian culture opened a space that allowed Andalusia to be interpreted not as a culturally backward region in a country that was on the margin of Europe but as an alternative to a now problematized Faustian/Northern European culture.

A second link between the works of Spengler and Lorca relates to the symbolism of the cave, where, according to the former, Magian culture and nature meet. Spengler describes the Magian imaging of the world as "cavernlike" (*DW,* 299), which he contrasts with the Faustian tendency to imagine the world in limitless space. For this reason Magian architecture, Spengler explains, tends to be cavernlike: "There is something in the basilicas of Christianity, Hellenistic, Hebrew and Baal cults, and in the Mithraeum, the Mazdaist fire-temple and the Mosque, that tells of a like spirituality: it is the *Cavern-*feeling" (*DW,* 111).[11] Frank, in *Virgin Spain,* also focuses on the figure of the cave, beginning with a description of mosque architecture in North Africa and moving to the description of contemporary cave homes in eastern Andalusia, particularly in Guadix in eastern Granada and villages in Murcia (*VS,* 45–47). Frank then proceeds to the city of Granada and focuses on the Sacromonte, and the caves where Romany families live. Lorca's evocation of caves as dwelling places—for example in "Cave," from *Poem of the Deep Song* (*OC,* 305) and in the theatrical work *Blood Wedding* (*OC,* 1225), as well as of voices and songs that come from the depths of one's being in *Poem of the Deep Song,* are suggestive as well of a cavern-feeling, to use Spengler's term. Lorca also uses poetic images that create a metaphorical dome effect, with such verses as "The field of olive trees opens and closes like a fan" and "The ellipse of a cry, goes from mountain to mountain" ("Poem of the Gypsy Siguiriya," *OC,* 296–97); "Under the bow of the sky / over its clean flatland / she shoots the constant / arrow of her river" ("Sevilla," *OC,* 308–9) and "The seven maidens sing / (Over the sky, an arch [bow] of sunset examples)" from "Song of the Seven Maidens," *Songs.*

Spengler's view of culture as inextricably tied to the land where it originates also provides a framework for our understanding of Lorca's portrayals of Andalusia.[12] The idea of a close relationship between culture and the land and the harmony between the city as cultural center and the realm of nature appears in much of Lorca's work, particularly in *Gypsy Ballads, Poem of the Deep Song, Songs,* and *Divan of the Tamarit,* where city, town, and countryside often are situated in a continuous space. Indeed, what appears in these poems is, perhaps, best described as "landscape," that is, a blending of nature and culture. Plants featured in this poetry are generally cultivated, as evidenced by the flowers and fruit trees such as lemon and orange trees, olive trees, and their respective fruits. Movement through space tends to occur on preestablished routes; it is channeled along streams and rivers, streets in town, and roads or paths in the countryside. The cosmos evoked by Lorca in these works creates the impression of a limited space, which Spengler associates with Magian culture. In Lorca's poetry the low point is often the sea, while height is suggested by a mountain. The moon and stars often preside over earthly events and appear in contact with the earth, not in a distant, limitless outer space. The poetic references to an arched firmament, described in the previous paragraph, reinforce the idea of a limited, enclosed cosmos. The time portrayed is present, contemporary, yet pre or simply nonindustrial. Lorca's portrayal of the countryside in the U.S. in *Poet in New York,* on the other hand, conforms to Spengler's description of the Faustian soul, with its emphasis on solitude, loneliness, and a sense of infinite space, while the American city, whose model for Lorca is New York, is alienating, inhuman in its capitalism and industrialism, and seeks to invade the divine space of the heavens with its skyscrapers, in keeping with Spengler's view of Faustian architecture as defying nature by rising into the sky. Lorca's Andalusia is "un-Western" in that it does not feature the problematization of the relationship between civilization and nature that characterizes much of the literature of northwestern Europe and North America, which Spengler has described as "Faustian."

A third similarity between Spengler and Lorca's thought relates to their focus on the figure and the symbolism of the arabesque. Just as Spengler sees the cavern as the prime model for Magian architecture, he selects the figure of the arabesque as the single most important element of the Magian tradition of plastic arts: "in the Arabian world, after Justinian and Chosroes-Nushirvan, Arabesque dissolved all the forms of architecture, painting and sculpture into style-impressions that nowadays we should consider craft-art" (*DW,* 109). He lists the arabesque among the principle elements of the Magian culture, along with alchemy and the Kabbala, the Philosopher's Stone, the Holy Scriptures, the inner form of the tales of the *Thousand and One Nights,* and the use of the gold as a background in artwork (*DW,* 130).

With regard to the arabesque, Frank once again takes on a mediating role, appearing to take a Spenglerian idea and apply it to Spanish, and primarily Andalusian, culture. He uses the term to describe the movement of the castanets in the gypsy dances he observes (*VS*, 84), and the movement of the body, "skeletal arabesque," in Andalusian dance (*VS*, 62). At the end of his study, he applies the image of the arabesque to the painting of Pablo Picasso and the poetry of Jiménez in a way that is reminiscent of Spengler's tendency to view the arabesque as the essential principle of Magian art:

> The Idea in Picasso is the *arabesque*. The moslem needed to harmonize his love of beauty and his fear of idols. He required a plastic form which would not, like the forms of physical nature, recall the old idolatries of the desert. The letters of his holy Arabic language served him, even as the Hebrew letters served, in the Kabbala, to make a Temple for the medieval Jews. These letters of the Arab held Allah and all the world of Islam, and yet were freed of natural associations. In Picasso, there is a similar impulse. His need also, as an artist, is to make designs; but his fear was an idolatry of another sort. The associations attached to physical forms—religion, sentiment, moral value—were already so unconvincing in the nineteenth century that Cézanne refused them, as Moses the Golden Calf. Now, even the sensory organisms formed by the eye must be rejected. (*VS*, 288–89)

Frank clearly views the arabesque not as an element of a decadent culture but as a kind of solution to a contemporary aesthetic dilemma. He establishes a link between the nonmimetic restrictions of Islamic and Jewish traditions and the antimimetic tendency of the most modern art, and in this case, the same tendency in a contemporary Spanish artist. Frank suggests that Picasso's art, itself, draws from an ancient "oriental" tradition:

> This formal use of transfigured substances is old in Spain. Spain's wildest excess of sculpture and ornament in the plateresque holds an element of transformed abstraction that recalls Egypt. The Spanish dance resolves dramatic gesture into a formal end. And now, Picasso makes an arabesque of the letters and signs of nature. (*VS*, 288–89)

Frank goes on to apply the idea or image of the arabesque to the poetry of Jiménez, another Andalusian poet, whose

> poems have prosodic value. Yet their chiefest value is that they *create aesthetically* a sense of incompleteness. Aesthetically, they are whole because they contain this *lack*—this positive surge toward an apocalyptic sense which lives in them only by the imprint of its absence. Each of his poems is at once a sensory form, and a spiritual inchoation. . . .
>
> No Castilian poet since Luis de Góngora has equaled Jiménez in craft and virtuosic power . . . Góngora worked as if in minerals. His arabesques

> were cut and carved, like the original arabesques, in stone. In Jiménez, the arabesque is of organic substance: it is traced in flesh, blood, breath . . . Jiménez has lifted from life his overtones into a form that is life's natural emergence into consciousness from the eternal infraconscious flow. His arabesques are therefore as organic as their base. Jiménez belongs to the race of Góngora and San Juan de la Cruz—a race of poets who are immortal, and hermetic." (*VS*, 291–92)

With regard to Jiménez's poetry, Frank uses the term *arabesque* to describe three main aspects of Jiménez's poetry: its sense of incompleteness, its emergence from the unconscious to consciousness, and its organic quality. While the question of Picasso's and Jiménez's cultivation of the arabesque in their work may be subject to debate, it is evident that Frank saw the arabesque as encapsulating some important aspect of a Hispanic tradition rooted in the East, a tradition which he portrays as not being at odds with itself, unlike that of the hegemonic European civilization.

Returning to Lorca, but keeping in mind the significance the arabesque had for a critic like Frank, one cannot fail to recognize the importance of the arabesque in Lorca's own creative work, as well as in his critical commentaries. Cecilia Cavanaugh's study of the arabesque in Lorca's drawings and poems offers insights into the combination of writing and image-making that Frank associates with Picasso's painting.[13] The nonmimetic quality of the arabesque would conform to Lorca's aesthetic objectives, as well as to those of Picasso and the avant-garde in general. In addition, Lorca may well have perceived the metaphorical possibilities of the arabesque as a transformative process, as Cavanaugh argues convincingly in her comparison between Lorca's critique of metaphor and Richard Klein's study on the metaphorical nature of the arabesque (*LDP*, 163). Cavanaugh demonstrates how Lorca uses the arabesque as a device for a process of writing and also of reading a poem, a process that involves both continuity and transformation across space and time (*LDP*, 166–69).[14] This transformative process in combination with the vegetal motifs that recur throughout *Divan of the Tamarit* suggest that Cavanaugh is correct in noting the presence of the arabesque, even though Lorca avoids using the word itself in the work. The fact that Cavanaugh has identified the most sophisticated use of the arabesque in Lorca's work as being after 1932 suggests that it relates to his mature understanding of his professional as well as of his personal identity. At the same time, it is clear that the arabesque also figures in Lorca's earlier work, as a commentary by Dalí on Lorca's drawings as early as 1927 indicates:

> The poetic system of Lorca's drawings tends toward an organic immateriality. Presiding over it is the finest, most physiological sort of calligraphy. Lorca, who is totally Andalusian, has a very ancient sense of how to relate color and architectonics. This sense is based on the uncontrolled, harmonious asymmetry that characterizes all the purest plastic creations of the Orient . . . At its best, Lorca's drawing [the Water Drop] partakes of the graphic life of certain lines drawn by the Surrealists and of the foolish, iridescent decorativism of the spiralling, colorful innard of crystal balls.[15]

Dalí's reference to Lorca's Andalusian orientalism is reminiscent of Spengler's association of Andalusia with Magian feeling. Moreover, Dalí's perception of "organic immateriality" and of a kind of calligraphy in Lorca's drawings also suggests the image of the arabesque, much as Frank described the figure in relation to the work of Picasso and Jiménez. Dalí's mention of calligraphy calls attention, once again, to a special correspondence between writing and drawing in Lorca's work, something that can be seen particularly in Lorca's playful use of the arabesque as ornamentation for his own name.

The fact that the arabesque may relate to music as well as to the plastic arts would not have been lost on Lorca, whose first love had been music. A kind of arabesque can be seen in Lorca's description of the *cante jondo*'s [deep song's] melody as "rica de giros ornamentales (lo mismo que en los cantos de la India), solo se emplean en determinados momentos como expresiones o arrebatos sugeridos por la fuerza emotiva del texto, y hay que considerarlos como amplias inflexiones vocales más que como giros de ornamentacion" [rich in ornamental turns (just as in the songs of India), used only at certain moments as expressions or raptures suggested by the emotive force of the text; it is necessary to consider them as broad vocal inflections more than as ornamental turns] (*OC*, 59). The poet features it in his theory of poetry and music, once again, relating Andalusian poetry and music, in particular, to a broader cultural context described by Spengler as "Magian."[16]

The movement implicit in the arabesque provides a connection to the next parallel between Spengler and Lorca, a point that relates to the term *Magian,* itself. In his introduction to *Decline of the West,* H. Stuart Hughes attributes the choice of the term *Magian* to Spengler's belief that the peoples so described "dwelt in a magic world of mysterious presences" (*DW,* ix). Certain literary works by Lorca feature atmospheres that correspond to a world of mysterious presences, especially *Poem of the Deep Song, Gypsy Ballads,* and *Blood Wedding,* works that evoke the Gypsy experience. At the same time,

Lorca, as critic, formulates a kind of theory of "mysterious presences" in his "Theory and Play of *Duende*," in which, moreover, the concept of *duende* serves to relate the image of the cavern to that of the arabesque. Near the beginning of his talk on *duende*, Lorca attributes the definition of the term to Goethe, quoting the German writer's reference to the music of Paganini: "Mysterious power felt by everyone and which no philosopher can explain" (*OC*, 110). Lorca then goes on to link *duende* to Andalusian culture:

> This "mysterious power felt by all and which no philosopher can explain" is, in summary, the spirit of the earth, the same *duende* that embraced the heart of Nietzsche, who sought it in its external forms on the Rialto bridge or in the music of Bizet, without finding it and without knowing that the *duende* he was pursuing had leaped from the mysterious Greeks to the dancers of Cadiz or to the Dionysian throat-splitting cry of Silverio's *siguiriya.* (*OC*, 110)

Lorca distinguishes *duende* from other spirits associated with artistic creation—"angel," or the angelic, and the muse:

> Angel and muse come from without; the angel gives light and the muse gives forms (Hesiod learned from them). Be it a golden beard or a fold of tunic, the poet receives norms in his little forest of laurels. On the other hand, it is necessary to awaken *duende* in the last chambers of the blood.(*OC*, 111)

Whereas he considers that the angel and the muse come to the artist from without, Lorca sees *duende* as inhabiting the depths of the person and coming forth, from the cavern of the body, so to speak, in song or dance. (Once again, this is unlike the Faustian "force" or "will" described by Spengler as a quality of Western art.) While Lorca affirms that all art forms and all nations may have access to *duende*, angel, and muse, he sees German art most related to the muse, Italian art related to the angelic, and Spanish art related most to *duende*, a spirit associated with movement. The poet sees in Spanish music and dance the most authentic manifestation of the culture, a culture open to death in a way other European countries are not (*OC*, 115). Also, while the images of an angel or a muse suggest the presence of a body, the *duende* seems to be a bodiless spirit:

> Ladies and gentlemen: I have erected three arches and with a clumsy hand I have placed in each the muse, the angel and the *duende*.
>
> The muse remains silent; she may have a tunic of small folds or the cow-eyes that in Pompey look at the large-nosed, four-faced figure her

> great friend Picasso has painted of her. The angel can shake the hair of Antonello of Messina, a tunic by Lippi and a violin of Massolino or Rousseau.
>
> The *duende* . . . Where is the *duende*? Through the empty arch enters a mental air that blows insistently over the heads of the dead, in search of new landscapes and unknown accents, an air with the smell of a child's saliva, of pounded grass, and a veil of Medusa that announces the constant baptism of things newly created. (*OC*, 120–21)

Lorca's idea of *duende* effectively dovetails with Spengler's view of Magian culture. As a generally bodiless spirit, it, much like the arabesque, corresponds to Spengler's belief that Magian culture tends to find ways to transcend the body or to make it abstract.[17] *Duende* is associated symbolically with air, or breath, and with blood, both symbols of life. At the same time *duende* is related to a wounding of the body and to the proximity of death. In this sense, it seems to belong to that liminal area where life and death meet.

In "Arquitectura del cante jondo" (Architecture of the deep song) Lorca describes the *cante jondo* in terms that suggest both the cavern-feeling and the arabesque of Spengler, while evoking the power of *duende*. His analysis draws on the research of Manuel de Falla into the Eastern roots of this Andalusian musical expression:

> The gypsy *seguidilla* begins with a terrible cry. A cry that divides the landscape in two equal hemispheres; then the voice stops to give way to an impressive and measured silence. A silence in which shines the face of a hot lily that has abandoned voice for the heavens. Next begins the undulating and unending melody in a different direction from that of Bach. The infinite melody of Bach is round, the phrase could repeat itself eternally in a circular course; but the melody of the *seguidilla* gets lost in a horizontal course, it escapes from our hands and we see it move away toward a point of shared aspiration and perfect passion where the soul does not manage to disembark. (*OC*, 57)

While Lorca describes the "infinite and unending melody" of Bach as circular, that of the seguidilla, or *siguiriya*, has a horizontal movement which suggests a combination of curves and straight lines. The repetition implicit in the arabesque appears as well in the repetition of the music, as Lorca's comparison of the *cante jondo* and Indian music indicates. Both musical forms feature "inharmoniousness as a modulating medium; the use of the melodic atmosphere, which rarely goes beyond the limits of a sixth, and the reiterated and even obsessive use of a single note, a procedure common to certain formulas of enchantment" (*OC*, 58).

Lorca also takes pains to show that the *cante jondo* is not merely a creation of the Gypsies:

> We are dealing with a completely Andalusian song which existed in seed-form before the arrival of the Gypsies, just as the horseshoe arch existed before the Arabs utilized it as a characteristic form of their architecture. A song that had already arisen in Andalusia . . . molded with the blood of North Africa, and probably with the deep, rending rhythms of Jewish peoples, the contemporary fathers of all the great Slavic music. (*OC*, 58)

Lorca's view of Roma artistic forms as one component of a broader and equally ancient Andalusian culture that is also in a sense Mediterranean is consonant with Spengler's understanding of Magian cultures that include various cultural and spiritual heritages: the Arabic, the Eastern Christian Church, Judaism, and those of more ancient peoples of the Middle East and North Africa.[18]

Spengler's belief that the science most developed by Magian cultures was alchemy provides a final link with the poetics of Lorca. Once again, Spengler considers alchemy to be intertwined with the cavern-feeling and the arabesque in the Magian worldview:

> The Arabian Culture can be symbolized by the arabesque and the cavern-vaulting of the mosque, and out of this world-feeling there issued Alchemy with its ideas of mysterious substances like the "philosophical mercury," which is neither a material nor a property but by magic can transmute one metal into another. (*DW*, 191)

In contrast with the physics of classical antiquity, concerned with matter and form, and with modern Faustian physics, occupied with force and mass, Spengler describes Arabian physics, much like the thought of Spinoza, as dealing with "*substances* with visible or secret *attributes*" (*DW*, 191).

Lorca's description of the *cante jondo* also incorporates a sort of alchemical power of transformation:

> I have spoken of "the voice of good blood" because the first thing that is needed for the song and the guitar-playing is that capacity for transformation and purification of melody and rhythm possessed by the Andalusian, especially by the Gypsy. A wisdom for eliminating the new and accessorial, so that the essential may emerge; a magical power to be able to draw or measure a *siguiriya* with an accent that is absolutely millenary. (*OC*, 60)

On one occasion Lorca's description of the *cante jondo* joins together the images of the cave or vaulted ceiling as firmament, the arabesque, and alchemy, much like Spengler had done in referring to the Magian worldview, artistic consciousness, and science:

> The figure of the singer [*cantaor*] is within two great lines: the arch of the sky in the external world, and the zig-zag that ascends within his soul.
>
> The singer, when he sings, celebrates a solemn ritual; he brings out the old, dormant essences and launches them into the wind wrapped in his voice. (*OC*, 60)

In the same text Lorca later describes the *siguiriya* as "a cautery that burns the throat and the tongue of the one who says it" (*OC*, 61), reiterating the reference to alchemy as a process for modifying or transforming substances, often by a heating process.

The transformation of one substance into another, the effect of the alchemist's work, parallels the task of the poet in his work with language, metaphor, and poetic images. In this sense, just as Lorca employed the arabesque in his drawing and in his poetic thought, he may have cultivated the idea of the poet as alchemist, for his references to metals and objects made of metal in *Gypsy Ballads* call attention to a process of transformation of metals, the stuff of alchemy. Unlike the traditional alchemist, however, Lorca's most prized metal is not gold, symbol of the eternal, but the silver of the ever-changing moon. His inclusion of metals and chemical substances such as nickel, arsenic, iodine, and silver, all possessing tones of a metallic gray, in addition to sulphur, in "Llanto por la muerte de Ignacio Sánchez Mejías" contributes to the "alchemical" overtones of the poem. According to Juan García Font, arsenic and sulphur are elements in the transformation of metals to gold; nickel and silver are used, and sometimes fused, in the process of coinmaking; some alchemical procedures called for the ingredient of human blood; (in Lorca's poem the blood of Sánchez Mejías is spilled and mixed with the sand); the preeminence of white, red, black, and yellow in Lorca's palette is also suggestive of alchemy, as these may symbolize four stages in the process of the transformation of substances. The poet's anguished description of the decomposition of Ignacio's body also corresponds to the chemical process of transformation, for "putrifaction" is one of the stages of the alchemical procedures.[19] Lorca's alternate references to extremes of heat and cold, and his bringing to bear the elements of earth, air, fire, and water in the poem, also suggest the transformation of matter implicit in alchemy, which Spengler associates with Magian cultures.

In conclusion, I believe that Spengler's theory of Magian culture provides a background that allows us to better understand Lorca's portrayal of Andalusia throughout his literary and critical writings. Lorca's portrayal of Andalusia's ethnic heritage, with its emphasis on the contribution of Arabic and North African, Jewish and Roma communities, his predilection for representing space as a domed or cavelike structure, his portrayal of the essence of Andalusian culture as arabesque or as *duende,* and the occasional chemical or "alchemical" resonance of his poetry, all are in consonance with Spengler's description of Magian culture. The fact that Lorca imbues his stylized Andalusia with positive aesthetic values suggests that he saw Andalusian culture as an alternative to the Faustian cultural model Spengler associated with northern Europe and the United States, even though Lorca was also critical of a number of traditional Spanish and Andalusian attitudes and customs. Lorca's interpretation of Andalusian culture was not, therefore, an anomaly in his otherwise progressive and at times revolutionary aesthetic vision. I suggest that he responds indirectly to the criticism of reformists who long had lamented the "non-Western" nature of Spain, and particularly Andalusia, by creating positive aesthetic values for what might be perceived as the non-Western, Magian, or oriental aspects of Spanish culture. A reading of Federico García Lorca's work in light of the cultural theory of Oswald Spengler and the insights of Waldo Frank substantiates the argument for the coherence of the Andalusian poet's ideas on his culture and its potential role in contemporary art and literature.

Notes

1. Even before the publication of *Gypsy Ballads,* Lorca expresses his discontent with the image being created of him. In a letter to Jorge Guillén he writes: "My myth of gypsydom is bothering me a little. They confuse my life and my character. I don't want that at all. The gypsies are a theme. And nothing more. I could just as well be a poet of sewing needles or hydraulic landscapes. Besides, the gypsy identity implies that I lack culture, education, and am a 'primitive poet,' which you know I am not. I do not want them to label me. I feel like they are putting chains on me." In Federico García Lorca. *Obras completas,* 6th ed. (Madrid: Aguilar, 1963), 1614. Hereafter *OC,* cited in the text.

In a letter to José Bergamín in 1927, Lorca reiterates his concern about his image: "Let's see if this year we get together and you stop considering me as a *gypsy;* you don't know how harmful that myth is to me and how *false* it is in essence, although it may not seem so in terms of form" (*Obras completas,* 1662). After the publication of *Gypsy Ballads* Lorca explains, referring to himself in the third person: "The gypsy identity is only one of many of the poet's themes; but it is not fundamental to his work, nor is it a constant theme. *Gypsy Ballads* is a book in which the poet has been successful in rendering the tone of the ballad and in dealing with a theme of his native land; but one cannot classify this poet exclusively as a poet of ethnicity, for his ambition is broader" (*Obras completas,* 1698). All quotes from Lorca are taken from this edition. Translations from the Spanish are my own.

2. References to Spengler's work are taken from Oswald Spengler, *The Decline of the West*, abridged edition, trans. Charles Francis Atkinson (Oxford: Oxford University Press, 1991). Hereafter *DW*, cited in the text. My study takes into account as well the Spanish translation of Spengler's work (see note 4, below).

3. Lorca's view of Andalusia as an alternative to "Faustian" culture is reminiscent of Miguel de Unamuno's call for an inward look toward the Hispanic and, specifically, Castilian tradition and values rather than seeking models in Northern Europe. At the same time, Lorca's affirmation of Andalusian culture may be viewed as a reaction against the tendency of the Generation of 1898 to seek the essence of Spanish culture in Castile, as Herbert Ramsden argues in the introduction to his edition of *Romancero gitano*. See Federico García Lorca, *Romancero gitano*, ed. Herbert Ramsden (Manchester: Manchester University Press: 1988), 1–18.

4. In his preface to the Spanish edition of *La decadencia de occidente*, 2 vols. (Madrid: Espasa Calpe, 1998), José Ortega y Gasset affirmed: "Oswald Spengler's book, *The Decline of the West*, is undisputably the most clamorous intellectual occurrence of recent years" (1:14). Additional references to Ortega's assessment of *Decline of the West* are taken from this Spanish edition. Hereafter cited in the test as *DO*; my translations.

5. Justo Romero, *Falla. Discografía recomendada. Obra completa comentada* (Barcelona: Ediciones Península, 1999), 48, reports that, according to John Brand Trend, Falla believed that the ballet *El amor brujo* would best be performed in a gypsy cave in Granada's Sacromonte. This association between a piece evoking Andalusia and a cave dwelling in Granada recalls Spengler's view of the cavern as the prime architectural symbol for Magian cultures.

6. All quotes from Waldo Frank are taken from his *Virgin Spain: Scenes from the Spiritual Drama of a Great People* (New York: Boni & Liveright, 1926).Hereafter *VS*, cited in the text.

7. Paul J. Carter in his study of Waldo Frank, *Waldo Frank* (New York: Twayne, 1967), mentions that Frank had written an essay on Spengler (116). In light of Frank's intellectual formation and interests, it is inconceivable that he would not have been familiar with *Decline of the West.*

8. Frank recognizes the "Celtiberian base" of the Iberian peninsula, which subsequently received and to some exent absorbed the arrival of Greeks, Carthaginians, Romans, Celts, Germanic peoples, Jews, and various Muslim groups from Northern Africa to Asia (*Virgin Spain*, 51–56).

9. Ian Gibson, *Vida, pasión y muerte de Federico García Lorca (1898–1936)* (Barcelona: Plaza y Janés, 1998), 26.

10. Upon his return from New York, Lorca speaks of the work that would become *Poet in New York* as an encounter between his poetic world and that of New York, affirming that both worlds have in common some of the same peoples: "In the midst of both are the sad peoples of Africa and its surrounding area . . . Jews, Syrians, and Blacks (*Obras completas*, 1699). He goes on to say, "I believe that being from Granada leads me to a sympathetic understanding of the persecuted. Of the gypsy, the Black, the Jew . . . the Moor, that we all carry inside" (ibid., 1700). Lorca's emphasis on African American culture in his comments on New York is an extension of his tendency to identify Andalusia with North Africa, as in his remarks to Sebastian Gasch from Lanjarón, Granada in 1927: "The Moorish accent sounds in all the tongues of the people. There is a wind from Africa, whose mists are in plain view. There is no doubt that here exists a pattern of nostalgia that is anti-European, but that is not oriental. Andalusia." The poet goes on to stress the North African presence in Spain: "In the midst of Sierra Nevada, one is in the heart of the soul of Africa . . . Strange and Berber Andalusia" (ibid., 1659).

11. While the domes of Christian churches are evidence to at least a remnant of the Magian soul, the mosques of Islam represent the Magian cavern-feeling more consistently, according to Spengler: "in the Arabian world, Islam, the heir of Monophysite and Nestorian Christianity and of the Jews and the Persians, carried the development through to the end . . . Islamic

domical building followed Mazdaist and Nestorian along the same tracks to Shantung and to India. Mosques grew up in the Far West in Spain and Sicily" (*Decline of the West,* 113).

12. Spengler treats the relationship he perceives between race, culture, and landscape in three chapters: "Origin and Landscape: The Cosmic and the Microcosm," "Origin and Landscape: The Group of the Higher Cultures," and "Cities and Peoples" (*Decline of the West,* 226–67).

13. Cecilia J. Cavanaugh, *Lorca's Drawings and Poems* (Lewisburg, Penna.: Bucknell University Press, 1995). Hereafter *LDP,* cited in the test.

14. See also Sandra Nadaff, *Arabesque: Narrative Structure and the Aesthetics of Repetition in 1001 Nights* (Evanston, Ill.: Northwestern University Press, 1991), 115–16, who addresses the similarities between arabesque and metaphor in the narrative structure of the *Thousand and One Arabian Nights.*

15. Mario Hernández, *Line of Light and Shadow: The Drawings of Federico García Lorca,* trans. Christopher Maurer (Durham, N.C.: Duke University Press, 1991), 22.

16. Frank's interpretation of Andalusian music provides a bridge between the thought of Lorca and Spengler; Frank sees a parallel between the effect of the castanets in Andalusian dance and the arabesque:

> The castanets click their dry commentary. Purring, shrilling, silent, they are a gloss subtle as Talmud page, as Arabesque, upon this whirl of sculpted fire which is the dance of Spain. The castanet came, probably, with the gypsy from the ruined culture of Byzance. It was known to the Greeks and Romans. In the hands of the *gitana,* it is—as perhaps it was in the Dionysian rites—a heightened bloodbeat to the music: simple, single dimensioned, metronomic. Here it is an instrument subtle as a voice. It is a running, rhyming prose. Like the arabesque—a decoration evolved from the written language—it retains its intellectual power. (*Virgin Spain,* 84)

Frank's analysis of Andalusian dance ends with a return to the "essential" as well as organic quality of the arabesque: "The Spanish dance is organic and essential. It is the one great classic dance surviving in our modern world" (ibid., 88).

17. Frank also reiterates the association of a bodiless soul with "Magian," Andalusian culture in his aphoristic description of Granada's Albaicín: "Al-Baicín has a soul. What Al-Baicín lacks is body!" To be sure, he goes on to balance this affirmation by describing the Alhambra ("Al-Hambra"), on the opposite hill in Granada, as a "soulless" body (ibid., 69).

18. See Adolfo Salazar, *Conceptos fundamentales en la historia de la música* (Madrid: Editorial Alianza, 1997), 35, who portrays Manuel de Falla as an impressionistic composer, whose impressionism, like that of Béla Bartók of Hungary and Igor Stravinsky of Russia, featured the encounter of Eastern and Western musical traditions.

19. Juan García Font, *Historia de la alquimia en España* (Barcelona: Editorial Creación y Realización, 1995), 60, 58, 22, 85–86. Lorca's use of the colors red, yellow, black, and white also may reflect the classical images and rhetoric that are featured in the poem. Spengler identifies these as the colors of classical painting (*Decline of the West,* 128).

Lorca beside Artaud: Parallel Developments of a Modernist Mise-en-Scène

C. Christopher Soufas Jr.
Tulane University

ALTHOUGH there has been little suggestion that either Federico García Lorca or Antonin Artaud may have overtly influenced the other (though Lorca may have been aware of Artaud's writing), these dramatists develop quite similar ideas of theater during the early 1930s that advance the movement toward reform begun earlier in the century. Pioneering efforts by stage innovators such as Constantin Stanislavski, Luigi Pirandello, and Bertold Brecht, among others, that bring greater professionalism to the stage primarily by means of their advocacy of an expanded directorial function (that barely existed at the beginning of the twentieth century), are seconded more than a decade later by Artaud and Lorca with an intensified emphasis on performance/audience dynamics. Artaud's notion of a "theater of cruelty," which demands maximum audience involvement in the performance, finds a strong resonance in Lorca's dramatic texts and other writings. Although his exploration of stage/audience relationships is central in experimental plays like *El público* and *Comedia sin título,* Lorca also integrates these insights into his commercial theater. In spite of the fact that Artaud remains influential as a primary source even today, his own attempts at stage direction met with failure. Lorca, however, was able to achieve considerable stage authority during his lifetime by instituting strategies that to a considerable degree parallel and complement Artaudian concepts. Thus, it will be useful to examine Lorca beside Artaud in order to underscore his significant theoretical and practical contribution to modernist theater reform in Europe.

Although there is a tendency to look upon early reform efforts as largely homogeneous and complementary, there are significant differences among prominent earlier innovators and the slightly younger Artaud. Unlike his most important predecessors, Artaud did not enjoy a sustained association with an established theater company, and this may explain in part why his ideas are decidedly more theoretical and provocative. Stanislavski's contributions are only marginally theoretical since his great contribution is a system for training actors which marshals an actor's personal emotional repertoire gathered from actual experiences in order to use them during the performance to identify more intensely with the emotions ascribed to a stage character.[1] Such an innovation has the secondary effect of destabilizing the relationship between dramatic text and the performance since the efficacy of the performance now depends upon factors unrelated to the text. Since the actor's contribution is unstable and the source of emotional inspiration finite, each performance will necessarily be different. What the audience witnesses, therefore, is actually a hybrid of text and gesture inspired in an experience often unrelated to the words of the script. This development is fundamental to the emergence of performance-based approaches to theater and, eventually, to the privilege accorded mise-en-scène over written text.

While making expanded creative use of the stage by means of music, lighting, and costume effects, often in collaboration with other artists whose interpretations are at times antagonistic to the original script, Brecht's principal innovations are also centered in acting. Organizing his plays around a minimal dramatic unit called the *gestus* (the sum of which constitutes the stage performance), the principal goal becomes to contrive means to distance the audience from the visual suggestiveness of theater and the complacency and passivity inherent in the mere contemplation of the stage.[2] To discipline the audience, Brecht institutes what he terms an "alienation effect," the systematic interruption of realist illusion, as a constitutive aspect of his theater. This requires that the actor step outside the role in order to make the audience aware of more important issues.[3] The practice of theater thus generates fresh varieties of metatheater, grounded in the extratextual experience of the actor in Stanislavski and in Brecht the active creation of a purposeful distance between story and spectator. Pirandellos's theater also decisively embraces a metatheatrical model, the revelation of a deeper reality—the latent "humanity" in a character displaced from a conventional starting point—beneath surface appearances and social masks. Pirandello's innovations are the formal consequences of a theory of "humor" in which tragicomedy and not high tragedy becomes the most appropriate vehicle to communicate such intuitions and to challenge audiences to accept a theater model that intrudes upon the previously more comfortable domain of the spectator.[4]

Modernist theory and practice seeks a fundamental reinvention of theater by means of active strategies of disruption, the immediate effects of which are to discredit an all too comfortable nineteenth-century realism. Artaud's writing, which begins well after Pirandello and Stanislavski become prominent, and more than a decade after Brecht undertakes some of his most notable theater projects, shifts the emphasis from the body of the actor to the production as a totality. The actor's body and gestures remain important, yet more fully integrated with the other aspects of the production. Conceived much like a living painting, there is no empty space in Artaud's conception of mise-en-scène. In contrast to Brecht's requirement to interrupt the performance from time to time, Artaud conceives the performance as incessant interruption and irritation. The desire to liberate theater from its dependence on recited dialogue demands a thoroughgoing reconception of theater which accords primacy to mise-en-scène as an integrated totality. What distinguishes Artaud from his slightly earlier contemporaries, therefore, is precisely his "cruelty," an intense vision that leaves no room for passive or "amused" spectating.

Artaud's extended manifesto on theater, *The Theater and Its Double*, continuously reiterates the need to replace traditional theater with an entirely different concept, the stage hieroglyph, a hybrid super-sign that merges the separate functions of the visual and verbal to produce a new stage language that will "leave no portion of the stage space unutilized [and with] the sense of a new physical language, based upon signs and no longer upon words . . . These actors with their geometric robes seem to be animated hieroglyphs."[5] In sharp contrast to Brecht, script, actor, and stage decorations should not divide the stage but rather the stage should function as an extended hieroglyphic complex that embodies new epistemological conditions which the spectator is given no choice except to embrace. The theater of cruelty demands that the spectator receive and accept the director's vision as perhaps the most intense form of metatheater imaginable, as a primary reality. The hieroglyphic conception of theatrical signs is integral, therefore, to a renewed theatrical understanding. Indeed, integration, the elimination of the type of distance that was so important to Brecht, becomes primary in Artaud's conception of a reconfigured stage. Rather than the "four wall" arch proscenium model that had dominated theater for nearly a century, Artaud proposes its exact antithesis: "The hall will be enclosed by four walls, without any kind of ornament, and the public will be seated in the middle of the room, on the ground floor, on mobile chairs which will allow them to follow the spectacle which will take place all around them" (*TID,* 96). Although this idea was never actually put into practice, it nevertheless reveals the intensity of Artaud's commitment to erasing all barriers between audience and performance.

Such a desire is also embodied in Artaud's notion of the "Single Creator" who embodies the functions of both playwright and director-producer. Although Pirandello and Brecht had largely attained such a status, as did others, Artaud's notion differs significantly from the Brechtian model that welcomed collaboration and conflict within the production for the sake of the alienation effect. Artaud's Creator imposes the directorial vision on an audience. Similar to Patrice Pavis's critical notion of what the theater audience ideally witnesses in a performance, the director's metatext of the mise-en-scène,[6] Artaud seeks a performance environment that eliminates the separate functions of dramatist, director, actors, and audience in order to bring them together as intimately as possible, artistically yet also psychologically and even physically. The situation that Artaud would create as a consequence is the "theater of cruelty," the forced marriage between stage and public that affords the audience no choice except to witness the totality of the director's vision. "Cruelty" is the elimination of dramatic distance, that forbids the easy, comfortable, passively received entertainment "to be accomplished by a thorough involvement, a genuine enslavement of the attention" (*TID,* 92). Not merely involvement in psychological speculation about a character's circumstances, the theater of cruelty aims to "create a metaphysics of speech, gesture, and expression, in order to rescue [theater] from its servitude to psychology and 'human interest' " (*TID,* 90). The constant goal is ultimately

> a visual language of objects, movements, attitudes, and gestures, but on condition that their meanings, their physiognomies, their combinations be carried to the point of becoming signs . . . the theater must organize it into veritable hieroglyphs, with the help of characters and objects, and make use of their symbolism and interconnections in relation to all organs and on all levels. (*TID,* 90)

A reinvigorated theater demands the total sensual/intellectual involvement of the audience under conditions which the audience must accept. This is the fundamental basis for "cruelty" as Artaud uses the concept.

With such ambitious goals, it should not be surprising that Artaud's productions were not well received. Indeed, the principal effort to translate his insights into practice, his adaptation of Shelley's *The Cenci,* marks a significant failure from which his directorial career never fully recovers. As a theorist, however, Artaud's ideas remain among the most influential. As an exact contemporary, Lorca's thinking about theater is especially relevant, even though the evolution of his theater takes place under much different conditions. Chastened by the failure of his first play, *El maleficio de la mariposa* that closed after only three performances, Lorca's subsequent theater becomes the focus of a reexamination of the performance possibilities for theater in Spain. Although alluded to in earlier plays such as *La zapatera prodigiosa* and *Amor de don Perlimplín con Belisa en su jardín, El público* represents his most important meditation on the power of audiences and the consequent need for theater reform. It is from this equally "cruel" personal and theoretical awareness of the considerable authority of Spanish theater audiences to dictate theater practices that Lorca's mature art begins.

Lorca understands early in his career that he must confront what Artaud could not: the issue of what a theater audience will do when challenged with an uncomfortable dramatic situation that forces it to become more actively involved in the representation than it might wish. *El público* offers strong testimony, as Lorca recognizes in contemporary interviews, that under present theater circumstances in Spain, the agenda it outlines for theater is unstageable: "como [*El público*] . . . es muy punzante y generalmente nada honroso, pues los espectadores se levantarían indignados e impedirían que continuara la represeantación. Sí, mi pieza no es una obra para representarse" [since *The Public* is very caustic and not very respectable, the audience would become indignant and stop the performance. Yes, my piece is not a work that can be staged].[7] To make the conditions viable to produce plays of this nature will require a realignment of theatrical forces, off stage as well as on. Spanish stage conditions at this historical moment, however, are analogous—but in an ironic, reverse sense—to Artaud's idea of cruelty: the failure of theater to exercise its authority has made the stage little more than a projection of the audience's most comfortable wishes. This may be perhaps why Lorca also ironically posits that the ugliness associated with the homosexual themes of *El público* actually gives form to the audience's unconscious thoughts, a competing mise-en-scène that conflicts with conscious efforts to bring an innovative production to the stage: "porque [*El público*] es el espejo del público. Es decir, haciendo desfilar en escena los dramas propios que cada uno de los espectadores está pensando, mientras está mirando, muchas veces sin fijarse, la representación" [because *The Public* is the mirror of the audience. That is, by having file by on stage the personal drama that each one of the spectators is thinking, often unaware, as they watch the play] (*OC*, 3:557). In a much more self-conscious manner, this position anticipates Pavis's notion that the work of theater is the director's metatext of the mise-en-scène actualized on stage before an audience, one metatheatrical vision presented to another set of eyes. In this peculiar context, however, the audience brings to the theater a more powerful agenda which confronts a creative-destructive force on the opposite side of the stage which Lorca envisions as the ultimate antidote to this type of theatrical tyranny: "una fuerza oculta cuando ya el público no tuviera más remedio que atender, lleno de espíritu y subyugado por la acción" [a hidden force when the audience has no other choice except to heed, filled with spirit and dominated by the action] (*OC*, 2:665), a counterforce that also happens to epitomize the very conditions of Artaud's theater of cruelty.

In Spain, there is the additional impediment of an entrenched theatrical establishment dominated by acting companies and theater impresarios who pander almost exclusively to applause and profit rather than to art. As Lorca pointedly lectures to all Spanish actors in his "Homenaje a Lola Membrives": "El teatro ha perdido su autoridad porque día tras día se ha producido un gran desequilibrio entre arte y negocio" [The theater has lost its authority because day after day there continues a great imbalance between art and business] (*OC*, 3:452). The function of stage director in Spain hardly exists since staging decisions reside largely in the hands of acting companies, only a few of which evince even a degree of professionalism. This is one reason why Lorca insists that Spanish theater companies adopt what by the 1930s has already become a European norm: "Así, pues, para lograr autoridad en el teatro no hay solo que montar obras buenas, sino montarlas con un imprescindible director de escena. Esto lo deben ir sabiendo todas las compañías. Hacen falta directores de escena autorizados y documentados que transformen las obras y las interpreten con un estilo" [Thus, to achieve authority in the theater it is not enough to present good works but to do so with an absolutely necessary stage director. All the companies must begin to know this. Essential are certified and credentialed stage directors capable of transforming works of theater and of interpreting them with a personal style] (*OC*, 3:454). The theater lacks authority in a double sense: unprofessional actors control a stage in turn dominated by intolerant spectators who want a comfortable experience which reconfirms values brought with them to the playhouse. Under these conditions, plays like *El público* will remain simply a private vision. Like Artaud, Lorca conceives his task as one of imposing much greater stage discipline on actors and audience by, to use Artaud's words, "[conducting] in the physical domain an exploration of intense movement and precise emotional gesture which is equivalent on the psychological level to the most absolute and complete moral discipline" (*TID*, 114–15). Lorca's manipulation of the dead weight of conventional theatrical expectations, however, arguably presents a more difficult task in Spain than in Artaud's Paris since enlightened reform of theater had made fewer inroads by the mid-1930s.

Lorca's mature theater—which begins with *El público* and *Así que pasen cinco años* and culminates in *Bodas de sangre*, *Yerma*, and *La casa de Bernarda Alba* (along with *Doña Rosita, la soltera*)—is best understood as a response to the acutely perceived impasse between audience intolerance and intensified performance demands initially raised in *El público.* The tactical differences between Lorca and Artaud must not be interpreted as evidence of a lack of artistic kinship with regard to the fundamental tenets espoused in *The Theater and Its Double.* Rather than to assault his audience or to require it to confront a dramatic situation from which it cannot retreat (and acutely aware that audiences are fully capable of exiting the theater at any moment), Lorca accomplishes an almost identical "enslavement of the attention" by gentler, more subtle manipulations. While the rural settings of these plays are certainly relevant to the themes developed therein, from a directorial perspective such settings also offer perhaps the greatest opportunity to minimize the evident phenomenon of actor pandering to audiences for applause and attention. Actors portraying rural characters crippled by their social circumstances are themselves restricted in their range of movement and gesture on stage. Of greater significance in relation to audience influence and presence in the conventions of performance are the geographic and temporal dimensions of Lorca's commercial theater scripts: the constant oscillation between vast and only vaguely defined geographic distances in *Bodas de sangre*; the obsessive, expansive wanderings of Yerma over time and terrain in *Yerma*; the prominence of a vast temporal dimension in the otherwise stage-bound settings of *Doña Rosita, la soltera*; and the stifling absence of physical space in *La casa de Bernarda Alba.* Although *La casa de Bernarda Alba* is often cited as evidence of Lorca's renewed interest in conventional realism, this play actually marks the culmination of Lorca's innovative manipulation of the theatrical receptive circuit. Continuing until his last unfinished experimental piece, *Comedia sin título,* Lorca's unwavering goal of renewed theatrical authority is premised upon a reordering of traditional stage/audience relationships.

The unmistakable phenomenon in the commercial plays, underscored by oscillating geographies of expansive/restrictive physical spaces (*Bodas de sangre*) and/or temporally extended/confining contexts (*Yerma, Doña Rosita, la soltera*), is that such spaces and/or times cannot be fully represented on stage and thus can only be fully accommodated in the spectator's imagination/consciousness. Lorca acknowledges the negative consequences of such stage dynamics in the fifth tableau of *El público* as the audience's physical appropriation of the stage, the result of its outrage over the Director's casting choice for Juliet in Shakespeare's *Romeo and Juliet* as a boy of fifteen rather than the young girl called for by the script. Nevertheless, a minority of sympathetic theatergoers, students, lament this failed experiment: "Aquí está la gran equivocación de todos y por eso el teatro agoniza. El público no debe atravesar las sedas y los cartones que el poeta levanta en su dormitorio" [Here is everybody's great mistake and why the theater is dying. The public must not transgress the sets that the poet constructs in his private chambers] (*OC,* 2:653). Nonetheless, the audience's domain fully extends into the space of the stage and the psychological space of its creator. Theatrical space is not free or autonomous, as the Director's impotent challenge to audience viewing prerogatives and his subsequent theatrical/existential demise underscore. The evolution of Lorca's commercial theater, however, is strongly guided by these private meditations, and it is also at this point at which Lorca and Artaud most clearly parallel each other. For both, theater becomes the occasion to devise stage circumstances that supplant the audience's conventional, comfortable authority manifested concretely in its dominance of the field of vision and thus the space of the stage. Although in Lorca more as the consequence of a subtle, extended process than the frontal bombardment of the senses advocated by Artaud, the constant goal of theater, as the Director proclaims near the end of *El público,* is to overpower the audience and to relegate it to a separate space that allows the dramatist to reveal previously hidden or forbidden truths "cuando ya el público no tuviera más remedio que atender" [when the audience has no choice except to heed] (*OC,* 2:665). Lorca's agenda for his commercial theater is to give tangible form and space to what hitherto had been relegated to invisibility.

In the commercial theater, Lorca reclaims authority and discipline over his audience by means of a process which acknowledges and then reverses the privileged vantage point of the spectator with regard to the stage events, a phenomenon perhaps best described as one of greatly enhancing the spectator's vision at the outset only to deprive it of such a capacity by the play's conclusion. *Bodas de sangre,* in fact, seems not to be a stage play at all by the midpoint of act 1 since the audience can virtually project the course of dramatic events. By the end of the first act, and more appropriate to melodrama than to high tragedy, it is more than clear that Leonardo plans to ruin the Novio's plans to marry the Novia, which will also entail the spilling of blood proclaimed in the play's title. Indeed, Lorca makes it easy for even the most inattentive spectator to project precisely such a scenario. If this were not enough, the final scene of act 1 features an exaggerated melodramatic dialogue between the Novia and her maid in which the Novia is forced to confess explicitly, and also to the audience, that throughout this tableau Leonardo has been gazing into her bedroom window from horseback at the very moment that the wedding arrangements are being finalized in another part of the house. The Novio is thus effectively cuckolded before the fact. By the beginning of act 2, the viewing perspective of the spectator is more than simply omniscient; the audience has been fully empowered to project a dramatic trajectory that portends a bloody denouement. At this point, however, viewing relationships begin to change drastically. Instead of being a witness to the type of wedding the spectator has already imagined, the audience is immobilized, left behind at the Novia's cave at the exact point from which Leonardo watched her earlier, outside her bedroom window, as the wedding party travels ten leagues to and from the ceremony and thus back to the reception. What had earlier seemed to be an omniscient and mobile perspective enhanced by the context of a dramatic context that consciously alluded to vast spaces between the properties of the wedding couple diminishes quickly to a much reduced and stationary point of view. The rest of act 2 further diminishes the spectator's vantage point. The audience neither sees nor hears any of the events preceding the escape of the Novia and Leonardo. Indeed, on-stage dialogue throughout this tableau consists largely of peripheral remarks or chatter that has little to do with an unfolding drama. Throughout this time, the space of the stage is filled by wedding guests who saunter across the stage only to be surprised by cries at the end that the adulterous couple has fled the scene. In contrast to the spectator's earlier dominant perspective, in this act the stage progressively becomes bereft of its tangible and material contents. Rather than affirming the spectator's earlier imaginative scenario, the stage now confirms precisely the opposite: like the fleeing couple, the representation itself has begun to elude an audience that had earlier projected a predictable outcome.

As act 3 begins, that diminished perspective is reduced further, underscored by a completely darkened stage in the first tableau in which woodsmen in a completely incongruous setting—given the fact that the geography of this area had been consistently portrayed as almost a desert that will barely support plants—much less a humid undergrowth of forest, which the stage directions now indicate. The apparent climax continues the spectator's excision from direct visual participation in the representation, as the death of the Novio and Leonardo is merely alluded to by their off-stage cries of death. In the final tableau, a scenario exactly the opposite of the vast tracts of space alluded to in act 1 emerges. Space effectively collapses as the final scene takes place in a completely white room where "no habrá ni un gris, ni una sombra, ni siquiera lo preciso para la perspectiva" [there will not be a gray, nor a shadow, not even enough for perspective] (*OC,* 2:788). Mimetic three-dimensionality is never stable in a play in which space and visual perspective is from the outset a variable mental construct. More than to present a faithful portrayal of rural life, the play represents the overturning of conventional representational relationships. As much as an entertainment ostensibly portraying the vicissitudes of a disciplined way of rural life gone aberrant, at a more profound metadramatic level the play more importantly is about the disciplining of vision, the deliberate manipulation of spectator perspective that literally force the conditions of representation into the spectator's consciousness and away from the physical site of its portrayal. If Lorca perceives a forbidden play like *El público* to portray the audience's most intimate yet inexpressible thoughts, in *Bodas de sangre* he creates stage conditions and viewing relationships that forcibly wed script, scene, and vision in a single climactic plane in which the consequences of the audience's earlier extended vision are revisited in a scenario bereft completely of space. The play's most profound content, therefore, is the violent wedding of script and scene at the expense of the comfortable space previously occupied by the theater audience.

Similar disruptions to realist space occur in *Yerma* and *Doña Rosita, la soltera.* Here, however, spatial displacement is the consequence of the necessity for the audience to juxtapose largely static tableaus separated by ever greater expanses of time. Thus, these plays become fully representable only by means of the spectator's mental juxtaposition of disparate temporal moments which again in practical terms serve to force the relocation of the primary site of the representation to the spectator's imagination/consciousness.[8] The audience becomes the primary medium, another metatext of the mise-en-scène that competes with the playwright's vision. The audience is called upon to reconstitute the play mentally in order for the temporally disconnected scenes to make sense as a representation. This is especially evident in *Doña Rosita, la soltera* in which thirty years pass from initial scene to the sad conclusion. What initially seems to be the audience's privileged point of view in *Yerma* is abruptly and violently curtailed in Yerma's murder of her husband, Juan.

This murder, however, is also an act of excision of those elements, represented also by the community of spectators that has consistently reaffirmed her husband's position regarding the child and that has also prevented Yerma from understanding the depth of her tragedy. To represent the tragedy of Yerma, therefore, requires that the dominant discourse embodied in the cultural-social values and the property of her husband, expressed in the play in scenic, visual terms, be overthrown. Juan's murder is also his removal from the scene of Yerma's personal tragedy, conditions which had hitherto prevented her from giving concrete form to her grief. At a metadramatic level, Juan's murder depicts another act of elimination, the audience's from the space of the stage whose values it has long dominated. As the present play concludes, Yerma awaits another scenario to emerge, the hitherto unrepresentable tragedy that now begins to unfold. With the disappearance of Juan, the fertile lands he was responsible for bringing into fruition will become a wasteland (*yermo*), a more appropriate scene for a tragedy presided over by Yerma. As Yerma acknowledges the loss of all hope for a child, she nevertheless creates new possibilities for the portrayal of her grief, which may now proceed from the private and intimate perspective she had previously been precluded from relating. As much as for the desperate search for a means to become fertile, the play is more appropriately about the search for a means to represent a more profound drama that now begins to unfold. Yerma's final declaration is to indicate that, unburdened of the dream of the child, she will now be able to "descansar sin despertarme sobresaltada, para ver si la sangre me anuncia otra sangre nueva" [rest without waking afraid, to see if this blood heralds another new blood] (*OC*, 2:880). Part of this "new blood" is also a new theater, the possibilities of which are only beginning to become evident.

While seeming to herald a return to the standard four-wall realism of the arch proscenium stage, *La casa de Bernarda Alba* is actually the culmination of these earlier manipulations of audience perspective. What seems to be an almost exact reduplication of the nineteenth-century formula for play-making instead becomes an intensified means of achieving the very effects about which Artaud theorized yet failed to achieve in a commercially viable manner in his own productions. The most significant deviation from realist expectations is the pronounced theatrical consciousness of the play's characters. The initial words of Bernarda Alba, after the funeral guests have departed, could also very well be those of a stage director after a performance: "¡Andar a vuestras casas a criticar todo lo que habéis visto!" [Go home to criticize all you have seen!] (*OC,* 2:984). In fact, owing to the looming prospect of reduced economic circumstances in the household because Angustias is to inherit the lion's share of the estate and leave the house, Bernarda can no longer depend on economic means or the stability of her husband's name and their marriage to uphold her position in the community. To assure her continued viability within her closed society, Bernarda resorts to theatrical means in order to define the conditions under which an honorable social life will proceed with her as the new head of the household. Such conditions require that the house itself become the scene of a very unconventional type of theater in which the mise-en-scène consists in exactly the opposite of what an audience would expect at a performance: the absence of anything that could be expressed visually or verbally that would be considered untoward regarding the house, that is, Bernarda's "playhouse." Bernarda's stunningly unorthodox theatrical strategy is thus to contrive a theatrical scenario in which the "script" consists of silence and the scenario is bereft of all action. As Bernarda understands her new position, only the absence of all negative discourse about her house will assure her authority and status. Lorca conceives a double drama, one whose goal is to fulfill his protagonist's desire to maintain silence about the inner workings of the house while the other, which the theater audience witnesses, represents the practical, "backstage" efforts of Bernarda to assure that the primary scenario proceeds according to her will.

From the outset, however, it again becomes evident that the theater audience occupies a point of view that is much more informed than Bernarda's. Indeed, Bernarda's entries into the scene are primarily at moments of stress or crisis-disagreements among the daughters—when the facade of silence is threatened. As would a stage director to reposition the actors, she enters the scene only to quell disturbances that threaten her understanding of the necessary relationship between her house and the townspeople. The theater audience, however, from the outset knows exactly what is happening and cannot help but realize that Bernarda is most ineffective in achieving the "stage" conditions that she considers essential. The theater audience, however, does not witness what happens in a room of four walls but rather occupies a much more intimate point of view, as an integral part of the household itself in a position analogous to that of the daughters, or rather, the servants, that is, inside the house and privy to the inner workings yet not actual participants in Bernarda's progressively uncontrollable scene. From very early on, it becomes apparent that the house does not belong to Bernarda in any meaningful sense but in fact to another off-stage player who has been able to disturb the privacy and silence that Bernarda so covets. This, of course, is Pepe el Romano, who is to marry Angustias, yet who is also secretly involved with Adela. Yet what is not apparent is that the point of view of the audience is so omniscient and superior to that of the blinded Alba that effective "ownership" of the scene by this point lies in the gaze of the audience. It is the only medium capable of containing the contents of a private drama that at any moment will spill out into the public domain. What Lorca does, in effect, is not simply to make the situation more intimate by his choice of setting but rather to situate the drama, yet more intensely than in his earlier plays, in the imagination/consciousness of his audience. The site in which the phrase "la casa de Bernarda Alba" has meaning is precisely in the mind of the spectator of this drama, since the audience becomes fully invested as the "owner" or "manager" of this property. The audience thus occupies the role originally designated for Bernarda. Its omniscience is thus also threatened by the embodied yet invisible force referred to Pepe el Romano, whose sexual predations threaten to overturn the formula that Bernarda instituted for the continued stability of her house. When Pepe succeeds in destroying this balance as Adela commits suicide after being told that Bernarda has killed him, Bernarda's scenario is no longer viable. Whether or not her daughter has died a virgin as she proclaims in the play's final lines, a new discourse-scenario has already begun in the imaginations of the townspeople, whose theme is exactly the opposite of hers. Bernarda's house is diminished, her authority is eclipsed, and the town, which upon Bernarda's downfall loses what hitherto had been its most authoritative and condemnatory moral voice, is itself fundamentally changed.

A force more powerful than the physical presence of Pepe el Romano, however, is the ultimate victor in this struggle. The actual Pepe is an equal casualty in these negotiations in that the wedding between him and Angustias no longer remains a possibility. Everyone, including the seemingly invincible Pepe who comes "por el dinero" [for the money], loses and is substantively diminished. Also diminished in this final moment, however, is the theater audience itself whose omniscience remains intact only for as long as the "casa de Bernarda Alba" is a viable construct. The audience's comfortable and omniscient position is also eclipsed as a play that purports to be about a "house" becomes instead an occasion to introduce a new protagonist, an invisible actor who proves to be the truest and most destructive protagonist.

In relation to Artaud's notion that conceives of a radically new configuration of theater with the walls of the stage surrounding the audience, Lorca accomplishes a stunningly similar effect here through more subtle if no less hostile means. Unbeknownst to the average theatergoer, he creates the precise conditions for "cruelty" in the Artaudian sense in that the audience has no choice except to function as the omniscient container of the contents of the house. The audience and not the stage "houses" this drama. Meaning is not produced on stage but rather only in the consciousness of the audience. Far more "cruel" than the character associated with dictatorial cruelty in the management of her house are the viewing relationships of the play. What destroys Bernarda and her daughters and diminishes the moral underpinnings of the townspeople is an invisible protagonist, the very force that the Director of *El público* proclaimed six years earlier that it was his mission to seek means to bring to light, "el perfil de una fuerza oculta" [the profile of a hidden force] (*OC,* 2:665). This is the authoritative protagonist that presides over the scenario of *La casa de Bernarda Alba.* A hidden force destroys everything and everybody, including its purported vehicle, Pepe. All fall in the house of Bernarda Alba that also doubles as the impotent site of Spanish theater practice, especially the theater audience which has significantly invested in its traditional yet ultimately untenable structure. The most profound message of the play is that there are no suitable mediums to contain such a force, which exactly coincides with the destructive insights of the unstageable *El público.* Lorca thus ironically gives form to a situation in which stable form is not possible. In doing so he creates a fundamentally "cruel" situation that corresponds both in spirit and letter to Artaud's theories as he continues to align his art with the radical theater sign that guided the vision of *El público.* Like Artaud, Lorca's dramatic goal is to put an end to "the subjugation of the theater to the text" in order "to recover the notion of a kind of unique language half-way between gesture and thought" (*TID,* 89). *La casa de Bernard a Alba* is not simply metatheater; rather it is a play within a play within yet another play directed and staged by an unseen protagonist which as it wreaks havoc on Bernarda's household also acquires the symbolic meaning that Artaud associates with the hieroglyph. It dislodges the theater audience from its comfortable bourgeois moorings and makes it impossible to remain outside the "four walls" of conventional theater. Putting his audience in a situation in which it is impossible to escape from consciousness, fundamental conditions for a theater of cruelty, Lorca forces that audience inside those walls in order to have it experience and partake of the same degradation and cruelty which his characters suffer.

As the Autor, the principal character of *Comedia sin título*, proclaims with words that could easily be attached to a prologue for *La casa de Bernarda Alba*: "el poeta os hace una encerrona porque quiere aspirar a conmover vuestros corazones enseñando las cosas que no queréis ver, gritando las simplísimas verdades que no queréis oír" [the poet shuts you inside here because he wants to move your hearts as he shows you things that you do not want to see, shouting the simplest of truths that you do not want to hear] (*OC*, 2:1069). Part of the Autor's vision for attaining such stage authority is the affirmation of a mode of play production very nearly identical to Artaud's idea of the "Single Creator." The type of authority that the Autor envisions is precisely one that centers responsibility in a single personality, one capable of telling "la verdad sobre los viejos escenarios" [the truth on the old scenarios] (*OC*, 2:1095). This is precisely what Lorca's mature commercial project dedicates itself to during the thirties, the exposure of truths that the theater-going public has no wish to see or hear, yet which nevertheless the playwright-director-producer dedicates his full being to achieving. In *La casa de Bernarda Alba*, the destructiveness of misplaced desire is identical to that of *El público*, the only difference being its physical "housing" in a different and more "comfortable" context. What Lorca accomplishes in his reordering of stage relationships is also the creation of a hieroglyphic supersign whose referent is ultimately not the visible scene but rather, as strongly alluded to in *La casa de Bernarda Alba*, a more intimate, personal drama that resists and defies representation as it paradoxically seeks ever more efficient means to transcend such impediments to its fuller presence. In this sense, Lorca fully embraces the essence of Artaud's ideas for the theater to transcend mere representation while he discovers what Artuad never could, a viable means to communicate such ideas before a theater audience. Seen in this light, Lorca's steadfastness to his theatrical ideals and his commitment to modernist reform remain strong throughout his career. While taking strategic but not substantive detours in his commercial plays, Lorca remains absolutely faithful to his creative-destructive vision in *El público*, which remains the foundation stone of his theater throughout his lifetime.

Notes

1. See Timothy J. Wiles, "Stanislavski's Theater of Affective Memory," *The Theater Event* (Chicago: University of Chicago Press, 1980), 13–36.

2. Bertold Brecht, *Brecht on Theatre,* trans. John Willet (New York: Hill & Wang, 1964), 33–42.

3. Ibid., 130–40.

4. Luigi Pirandello, *On Humor,* ed. and trans. Antonio Illiano and Daniel P. Testa (Chapel Hill: University of North Carolina Press, 1974), 107–45.

5. Antonin Artaud, *The Theatre and Its Double: Essays,* trans. Victor Corti (New York: Grove Press, 1958), 54. Hereafter *TID,* cited in the text.

6. Patrice Pavis, "From Text to Performance," *Performing Texts,* ed. Michael Isaacharoff and Robin F. Jones (Philadelphia: University of Pennsylvania Press, 1988), 86–100.

7. Federico García Lorca, *Obras completas,* vols. 2 and 3 (Madrid: Aguilar, 1986), 3:557. Hereafter *OC,* cited in the text. Translations from the Spanish are my own.

8. See Luis Fernández Cifuentes, *García Lorca en el teatro: La norma y la diferencia* (Saragossa: Prensas Universitarias de Zaragoza, 1986), 219.